Microbiology Monographs
Volume 10

Series Editor: Alexander Steinbüchel
Münster, Germany

Microbiology Monographs

Volumes published in the series

Preface

In 1898, an Austrian microbiologist Heinrich Winterberg made a curious observation: the number of microbial cells in his samples did not match the number of colonies formed on nutrient media (Winterberg 1898). About a decade later, J. Amann quantified this mismatch, which turned out to be surprisingly large, with non-growing cells outnumbering the cultivable ones almost 150 times (Amann 1911). These papers signify some of the earliest steps towards the discovery of an important phenomenon known today as the Great Plate Count Anomaly (Staley and Konopka 1985). Note how early in the history of microbiology these steps were taken. Detecting the Anomaly almost certainly required the Plate. If so, then the period from 1881 to 1887, the years when Robert Koch and Petri introduced their key inventions (Koch 1881; Petri 1887), sets the earliest boundary for the discovery, which is remarkably close to the 1898 observations by H. Winterberg. Celebrating its 111th anniversary, the Great Plate Count Anomaly today is arguably the oldest unresolved microbiological phenomenon.

In the years to follow, the Anomaly was repeatedly confirmed by all microbiologists who cared to compare the cell count in the inoculum to the colony count in the Petri dish (*cf.*, Cholodny 1929; Butkevich 1932; Butkevich and Butkevich 1936). By mid-century, the remarkable difference between the two counts became a universally recognized phenomenon, acknowledged by several classics of the time (Waksman and Hotchkiss 1937; ZoBell 1946; Jannasch and Jones 1959).

Surely the "missing" microbial diversity was as large then as it is now. However, reading the earlier papers leaves an impression that throughout most of the 20th century the "missing" aspect was not viewed as a particularly important problem or as an exciting opportunity. A casual mention was typical of many publications. "Missing" cells were not necessarily considered missing species let alone signs of novel classes of microbes. Besides, the unexplored microbial biodiversity was a purely academic issue; the hunt for novel species as a resource for biotechnology had not yet begun. It is also important that the reasons for the Anomaly appeared rather simple at the time. Counting errors, dead cells, and later damaged cells were continuously considered significant components of the disparity. Also, it had been obvious at least since Koch's time that no single nutrient medium could possibly satisfy all microorganisms (Koch 1881), and so the finger was always pointing to media deficiencies. Indeed, imperfections in media design was such a simple and intuitive explanation for the refusal of the microbial majority to grow *in vitro* that many microbiologists began viewing it as sufficient. The triviality of the explanation generated a perception of the Anomaly as a purely technical issue that could be resolved by bettering the media compositions and incubation conditions.

This view began to change towards the end of the 20th century. Cultivation efforts during the preceding decades did produce success stories; yet even as the manuals for media recipes grew into thick volumes, the overwhelming majority of microorganisms still eschewed the Petri dish. The progress in recovering missing species was rather incremental and did not change the overall picture. And, it was going to get worse.

The rRNA approach (Olsen et al. 1986) was a truly spectacular development: it provided insight into the microbial world missed by traditional cultivation. Novel microbial divisions were discovered by the dozen (Giovannoni et al. 1990; Ward et al. 1990; DeLong 1992; Fuhrman et al. 1992; Liesack and Stackebrandt 1992; Barns et al. 1994; Hugenholtz et al. 1998; Ravenschlag et al. 1999; Dojka et al. 2000). From the molecular surveys of the 1990s emerged an image of the biosphere with millions of novel microbial species waiting to be discovered (Tiedje 1994; Allsopp et al. 1995). What microbiologists had been able to cultivate and catalogue throughout the entire history of microbiological exploration (Staley et al. 1989) appeared to be an insignificant portion of the total. Successes in cultivation notwithstanding, the gap between microbial richness in nature and that of culture collections just would not close. Even today, most of the known microbial divisions have no single cultivable representative (Rappe and Giovannoni 2003; Schloss and Handelsman 2004). This gap was called "extraordinary" in 1932 just as it was called in 2000 (Butkevich 1932; Colwell 2000), as if the countless cultivation studies during these seventy years never existed. But, the realities of our age are different from the 1930s, and the Great Plate Count Anomaly is no longer "just" an academic observation. The need to close the gap is an urgent practical issue, as biotech and pharmaceutical industries appear to have exhausted what the limited number of cultivable species have to offer (Osburne et al. 2000). Today, the resolution of the phenomenon of microbial uncultivability is recognized as a top research priority for microbial biology (Young 1997; Hurst 2005). The principal challenges are to understand why uncultivated microorganisms are uncultivated, and to describe, access, and utilize their seemingly infinite diversity.

Microbiologists answered the call using two different strategies. One represents a group of clever approaches that bypass cultivation altogether. These go straight to the genes of the "missing" species to mine them for the information and products they encode, or employ isotopes and miniature electrodes to measure the activities of these species *in situ*. It is truly exciting to see how, today, cultivation-independent studies can be done at a single cell level. The other is a head-on strategy, and consists of a multitude of innovations in cultivation, principally aiming at mimicking natural conditions. The two strategies have their specific advantages and disadvantages, but few microbiologists think it is a battle of two competing products. Instead, the likely solution to the Anomaly is in a symbiosis between the two. What form and shape this symbiosis will take, it is too early to say, but the good news is that both the authors and the readers of this book will likely witness, and witness soon, the process and the conclusion of this evolution.

Furthering this unification is the main goal of this volume. The contributions center around three themes. The first theme groups together several chapters that focus on what can be learned about the microbial world without cultivating it. John Bunge opens the volume by describing how to statistically estimate the size of microbial diversity using gene sequence data. Chapters by Mitchell Sogin and Terry Gentry et al. provide an account of the state of the art in recovering the sequence data from environmental samples. Antje Boetius et al. offer a perspective on studying microbes in nature by measuring their biogeochemical activities. Mircea Podar et al. explore how much information modern genomics tools can recover from single cells of uncultivated species. The second theme is the nature of uncultivated microorganisms, why so many species remain uncultivated, and how to domesticate them in the lab. Thomas Schmidt and Allan Konopka dissect the nature of slow growing species. Rita Colwell describes cells that are viable but nonculturable. Slava Epstein attempts to build a general model of the Great Plate Count Anomaly. The third theme builds connections between the Anomaly and practice. Vivian Miao and Julian Davies explore how metagenomics approaches could help to provide access to bioactive compounds produced by uncultivated species. Kim Lewis advocates a connection between the phenomemon of microbial uncultivability and antimicrobial tolerance of biofilms and persister cells. Ken Nealson's chapter concludes the volume by discussing the application of uncultivated cells to the search of life outside our planet, so as to outline the wide spectrum of subjects connected to the Great Plate Count Anomaly.

There is something else that this volume intends to convey. Working in the field of uncultivated microorganisms today involves both luck and privilege. Not everyone has the fortune to study a phenomenon that has endured for a century, is of unquestionable importance, and yet remains unresolved. It is fascinating to think that "our" phenomenon predates the model of the atom, the theory of the Big Bang, cracking the genetic code… It is humbling to think of the great minds who have contributed over the past century to its resolution. And it is sensational to think how enormously beneficial this resolution may be by providing unimpeded access to the missing microbial diversity, and the treasures therein. As to the luck … the luck is in the timing, for the right of entry into the world of uncultivated microbes seems to be just round the corner.

Boston MA
January 2009

Slava S. Epstein

References

Allsopp D, Colwell RR, Hawksworth DL (1995) Microbial diversity and ecosystem function. CAB International, Wallingford, UK

Amann J (1911) Die direkte Zählung der Wasserbakterien mittels des Ultramikroskops. Centralbl f Bakteriol 29:381–384

Barns SM, Fundyga RE, Jeffries MW, Pace NR (1994) Remarkable archaeal diversity detected in a Yellowstone National Park hot spring environment. Proc Natl Acad Sci USA 91:1609–1613
Butkevich NV, Butkevich VS (1936) Multiplication of sea bacteria depending on the composition of the medium and on temperature. Microbiology 5:322–343
Butkevich VS (1932) Zür Methodik der bakterioloschen Meeresuntersuchungen und einige Angaben über die Verteilung der Bakterien im Wasser und in den Büden des Barents Meeres. Trans Oceanogr Inst Moscow 2:5–39 (in Russian with German summary)
Cholodny N (1929) Zur Methodik der quantitativen Erforschung des bakteriellen Planktons. Zentralbl Bakteriol Parasitenkd Infektionskr Hyg A 77:179–193
Colwell RR, Grimes DJ (2000) Nonculturable microorganisms in the environment. ASM Press, Washington DC
DeLong EF (1992) Archaea in coastal marine environments. Proc Natl Acad Sci USA 89:5685–5689
Dojka MA, Harris JK, Pace NR (2000) Expanding the known diversity and environmental distribution of an uncultured phylogenetic division of bacteria. Appl Environ Microbiol 66:1617–1621
Fuhrman JA, McCallum K, Davis AA (1992) Novel major archaebacterial group from marine plankton. Nature 356:148–149
Giovannoni SJ, Britschgi TB, Moyer CL, Field KG (1990) Genetic diversity in Sargasso Sea bacterioplankton. Nature 345:60–63
Hugenholtz P, Goebel BM, Pace NR (1998) Impact of culture-independent studies on the emerging phylogenetic view of bacterial diversity. J Bacteriol 180:4765–4774
Hurst CJ (2005) Divining the future of microbiology. ASM News 71:262–263
Jannasch HW, Jones GE (1959) Bacterial populations in seawater as determined by different methods of enumeration. Limnol Oceanogr 4:128–139
Koch R (1881) Zur Untersuchung von pathogenen Organismen. Mitth a d Kaiserl 1:1–48
Liesack W, Stackebrandt E (1992) Occurrence of novel groups of the domain Bacteria as revealed by analysis of genetic material isolated from an Australian terrestrial environment. J Bacteriol 174:5072–5078
Olsen GJ, Lane DJ, Giovannoni SJ, Pace NR, Stahl DA (1986) Microbial ecology and evolution: a ribosomal RNA approach. Annu Rev Microbiol 40:337–365
Osburne MS, Grossman TH, August PR, MacNeil IA (2000) Tapping into microbial diversity for natural products drug discovery. ASM News 66:411–417
Petri RJ (1887) Eine kleine Modification des Koch'schen Plattenverfahrens. Centralbl f Bakteriol 1:279–280
Rappe MS, Giovannoni SJ (2003) The uncultured microbial majority. Annu Rev Microbiol 57:369–394
Ravenschlag K, Sahm K, Pernthaler J, Amann R (1999) High bacterial diversity in permanently cold marine sediments. Appl Environ Microbiol 65:3982–3989
Schloss PD, Handelsman J (2004) Status of the microbial census. Microbiol Mol Biol Rev 68:686–691
Staley JT, Bryant MP, Pfennig N, Holt JG (1989) Bergey's manual of sytematic bacteriology
Staley JT, Konopka A (1985) Measurement of in situ activities of nonphotosynthetic microorganisms in aquatic and terrestrial habitats. Annu Rev Microbiol 39:321–346
Tiedje JM (1994) Microbial diversity: of value to whom? ASM News 60:524–525
Waksman SA, Hotchkiss M (1937) Viability of bacteria in sea water. J Bacteriol 33:389–400
Ward DM, Weller R, Bateson MM (1990) 16S rRNA sequences reveal numerous uncultured microorganisms in a natural community. Nature 345:63–65
Winterberg H (1898) Zur Methodik der Bakterienzahlung. Zeitschr f Hyg 29:75–93
Young P (1997) Major microbial diversity initiative recommended. ASM News 63:417421
ZoBell CE (1946) Marine microbiology: a monograph on hydrobacteriology. Chronica Botanica Co, Waltham, MA, USA

Contents

Statistical Estimation of Uncultivated Microbial Diversity

J. Bunge (✉)

Abstract The full microbial richness of a community, or even of an environmental sample, usually cannot be observed completely, but only estimated statistically. This estimation is typically based on observed count data, that is, the counts of the representatives of each species (or other taxonomic units) appearing in the sample or samples. "Abundance" data consists of counts of the numbers of individuals from various species in a single sample, while "incidence" (or multiple recapture) data consists of lists of species appearing in several or many samples. In this chapter we consider statistical estimation of the total richness, i.e., the total number of species, observed + unobserved, based on abundance or on incidence data. We discuss parametric and nonparametric methods, their underlying assumptions, and their advantages and disadvantages; computational implementations and software; and larger scientific issues such as the scope of applicability of the results of a given analysis. Some real-world examples from microbial studies are presented. Our discussion is intended to serve as an overview and an introduction to the literature and available software.

J. Bunge
Department of Statistical Science, Cornell University, Ithaca, NY 14853, USA
e-mail: jab18@cornell.edu

Microbiol Monogr, doi:10.1007/7171_2008_3

1 Introduction

Recent research has shown that microbial communities are astonishingly diverse; in fact many studies only capture a small fraction of the diversity of a given community, despite intensive sampling efforts (Huber et al. 2007). In such cases we must estimate the total diversity – observed plus unobserved – by statistical extrapolation from the available data. This is a nontrivial and indeed not entirely solved problem in statistics; it is a topic of considerable interest and activity among theoretical (mathematical) statisticians, and its literature continues to evolve at a rapid rate (Bunge and Barger 2008). Some of these statistical developments have entered the mainstream of microbial diversity research, but some have not. In this chapter we give an overview of the area from an applied, data-analytic perspective, with the goal of providing the practitioner with a conceptual framework for the diversity estimation problem; the types of data typically encountered; and the relevant statistical procedures that are applicable to such datasets.

First we require a definition of "diversity." This in turn requires that the community or population in question be classified in a clear and unambiguous manner, i.e., that it be subdivided into mutually exclusive subsets which, together, comprise the entire population. For statistical purposes any well-defined classification system will do, but in a biological population it is natural to classify individual organisms according to the Linnaean hierarchy, in particular by "species.' However, there is currently no consensus regarding the concept of "species" for microorganisms, and instead microbiologists often group environmental microorganisms into operational taxonomic units (OTUs) based on their rRNA gene sequence similarity (Stackebrandt and Goebel 1994). A species is then provisionally defined to be a group (OTU) of cells sharing a certain percentage identity of their 16s rRNA gene sequences. Values from 97 to 99% are typically used.

Given a classification system, several indices of diversity have been defined (Magurran 2004), but the simplest is the number of OTUs, or "species richness' in a given community. This quantity has a clear physical meaning and in principle could be determined exactly, given unlimited sampling effort. However, species richness, while relatively straightforward to define, is difficult to ascertain in practice, because biological communities often comprise a few large and many small species, and it is precisely the small species that elude sampling efforts. That is, the unobserved part of the community may be subdivided into many small groups unbeknown to us, yet we are required to estimate the number of these unobserved species. This is why the statistical problem of estimating species richness does not at present have an optimal, universal solution. Indeed, some authors have argued (mathematically) that no such global solution is possible, and that under the most general, nonparametric formulation of the problem one can at best provide a lower bound for the species richness of a given population (Mao and Lindsay 2007). On the other hand, if one is willing to impose certain structural constraints, richness estimation becomes possible, although subject to the validity of the assumed structure. For this and other reasons it is advisable to use and compare several existing methods, which make different assumptions about the (unknown) structure of the population.

The goal of this chapter is not to comprehensively review the current literature or practice (statistical or biological), but to describe the scope and applicability of the major statistical methods from a synoptic, and somewhat idealized, perspective. (In particular, the references given here are intended as entry points to the literature not definitive historical summaries.) This is because the status of current theory and practice are, to a certain degree, fragmented and incomplete. The various methods have not yet been unified in a single mathematical framework, and in particular there is no comprehensive expository textbook, at the theoretical or applied level. More importantly from the practitioner's point of view, there is no unified and comprehensive software program for species richness estimation. Some methods have been implemented in software that can be readily used by the applied practitioner, others in software that requires a statistical computing specialist, while for others no software exists at all. In this chapter we seek to give an overview of the state of the art. We take a broad perspective, attempting to look beyond the present limitations of the literature or software resources (which at any rate are being continually improved), while referring the reader to current and relevant existing resources where possible. We focus on those methods for which the mathematical foundations have been studied in depth.

Generally speaking, two types of data are encountered in species richness estimation: first, abundance or frequency count data, usually from a single sample; and second, incidence or occurrence data, usually from multiple samples (from the same community). In the next two sections we discuss statistical methods for each of these data types, and connections between them. In the final section, we discuss certain scientific issues (not purely statistical) and potential future directions.

2 Abundance Data

In this scenario we collect a sample of organisms, sort them into species, and count how many of each kind we have in the sample. Such a description hides the complications of the data-collection process, which may have several stages, each with its own biases (as in the case of clone library construction), and it hides the somewhat arbitrary decisions underlying the operative definition of species or OTU. However, the procedure is at least conceptually clear, and we will relegate its uncertainties to the background for now, in order to focus on the statistical methods.

Given such a sample, then, how can we interpret it statistically? Since the total species list is unknown (otherwise there would be no estimation problem), there is no obvious ordering of the species observed in the sample. We therefore organize the data by simply counting the number of species observed once (the "singletons"), twice, three times, and so on. For example, in the dataset (1,25), (2,7), (3,7), (4,4), (5,1), (6,2), (8,1), (11,1), (13,1), (14,1), (16,1), (27,1), (31,1), and (37,1); there were 25 species observed once (each), seven observed twice, seven observed three times, …, and 1 observed 37 times (example data from Behnke et al. (2008); OTUs defined at 98% sequence similarity level). Thus there were

25 + 7 + 7 + 4+...+1 = 54 species observed altogether in this sample, and there were 1 × 25 + 2 × 7 + 3 × 7 + ... + 37 × 1 = 250 individuals.

If we denote such a dataset in general by $\{(i,f_i), i = 1,2,\ldots\}$, then i is the *frequency* (of sample occurrence) and f_i is the *frequency of the frequency i* (in the sample), an unwieldy phrase which we may replace by *frequency count*, and we call this *frequency count data*. It has a simple and intuitive graphical display with frequency on the horizontal axis and count on the vertical axis, as shown for our example data in Fig. 1.

The left-hand side of the graph represents the less-abundant or rare species (at least in terms of their representation in the observed sample), and the right-hand side represents the abundant or frequent species (in the sample). Note that the structure shown in Fig. 1 is typical of microbial diversity studies: almost half of the observed species (25/54) were represented by singletons in the sample, but at the same time there were three highly abundant species (with frequencies 27, 31, and 37), each accounting for more than 10% of the sampled individuals (27/250, 31/250, 37/250). This structure reflects the statement noted in the Introduction, that the unobserved portion of the population may be subdivided to an almost arbitrary degree, rendering species-richness estimation difficult, or at least prone to statistical error. Nevertheless, it is remarkable that statistical methods can often achieve usable and credible results in such situations, although they must be interpreted with care. There are two main families of methods for abundance data, and we discuss these next.

2.1 Parametric Abundance Models

In this approach we assume that each species has a certain propensity to enter the sample. (This propensity is not identical to its literal abundance in the population, because the production of the ultimate sample may not transparently represent the underlying population.) We call this the "sampling intensity" of the species: it is the number of representatives of the species expected to enter the sample during

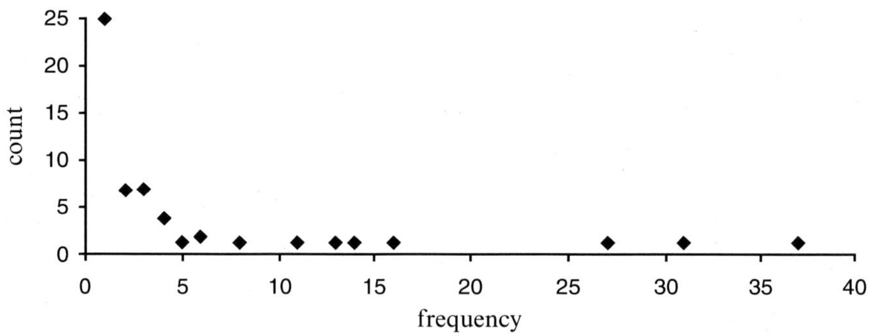

Fig. 1 Frequency count data example

one unit of sampling effort. The simplest possible assumption is that all sampling intensities are equal; however, this almost invariably results in underestimating the true species richness, often severely. It is more realistic to assume that the sampling intensities differ across species, some larger and some smaller. We then model the distribution of the intensities using a probability distribution which in turn is determined by a small number of parameters (for example, its mean and variance). This is a *parametric stochastic abundance model.*

If we suppose that each species independently contributes a random number of representatives to the sample according to the Poisson distribution, with mean equal to the species' sampling intensity, we have a *mixed-Poisson model* for the frequency counts (the "mixing" distribution is the distribution of the sampling intensities). We then fit the mixed-Poisson model to the observed frequency counts, generally via the method of maximum likelihood. (For an outline of the mathematical theory see Chao and Bunge 2002 and Bunge and Barger 2008). This amounts to fitting a "curve" to the data; the curve is projected upward and to the left to obtain an estimate of f_0, the number of unobserved species (i.e., observed zero times in the sample), and f_0 is added to the observed number of species to obtain a final estimate of the total richness. The same mathematical structure yields a standard error (SE) for the estimate, goodness-of-fit assessments, and so forth. Figure 2 shows such a model fitted to our example data. Here the estimate of f_0 is 67, for a total richness estimate of 54 + 67 = 121 species; the associated SE is 39, and the model fit is excellent. (Note that the curve is only fitted to the data up to frequency = 16; we discuss this later.)

Several questions arise immediately. First, how do we choose the parametric model, or stochastic abundance distribution, to use? It would be ideal if basic ecological theory would provide such a model, and it could be confirmed to fit data in a large number of cases. However, while there has been considerable work both in the mathematical and biological literature on the derivation of such models, no consensus has emerged (Williamson and Gaston 2005). Furthermore, it is not clear that a single

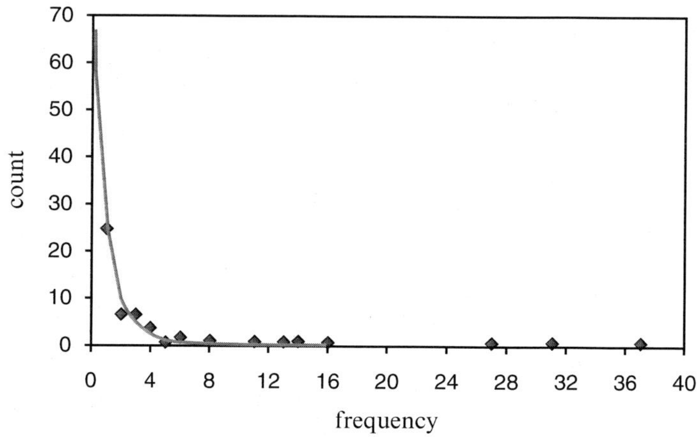

Fig. 2 Frequency count data example with fitted parametric model

model would apply across different types of organisms, or at different levels of the taxonomic hierarchy. Even if such a model could be found for the abundances of microbial species in a specific habitat, under specific conditions, it is not obvious that the model would remain unchanged under the various sampling, molecular biology and bioinformatic procedures that are used to construct a clone library, which is often the final source of data for analysis. One can envision a seamless theory that would mathematically describe environmental abundances, sampling, and the production of the clone library, resulting in a final model for the library as a representation of the underlying population; however, no such theory has yet been attempted. Hence at present the choice of model must generally be empirical, based on apparent fit to the data. We return to this question below.

Second, we typically fit models only to the observed frequency count data up to some maximum frequency, which is called the "right truncation point" or "tuning parameter," denoted by τ. Given an analysis based on the truncated data, we simply add the number of species with frequencies greater than τ to obtain the final richness estimate. (In the example above, $\tau = 16$.) The reason for this is that it may be impossible to find a parametric model (from any known family) that fits a given complete frequency-count dataset. Some researchers justify this truncation heuristically by noting that much of the information relevant to estimation of f_0 may be found in the lower frequencies (more rarely observed species), since these strongly influence the trajectory of the fitted curve, upward and to the left. However, the statistical foundation of this statement is uncertain, and in fact the richness estimate (and its SE) will typically vary with τ.

Nevertheless it is usually necessary to select a value for τ, and there are several possible approaches to this choice. First, we can look for an apparent gap in the data, and use that to subdivide it into "rare" (low-frequency) and "abundant" (high-frequency) species. In the example, such a gap occurs between 16 and 27, so $\tau = 16$ is not unreasonable. This approach requires expert judgment in each case, and furthermore there may be no such gap, or there may be many. Second, we can set τ at a heuristically-selected fixed value, say $\tau = 10$. But if τ can be taken much higher than 10 while still obtaining a good fit, potentially important data points will be omitted unnecessarily. Third, we can set τ to be the maximum value that will allow a good fit of a given parametric model; however, this value (and hence the results of the analysis) will vary depending on the model. Finally, we can take τ to be the maximum observed frequency in the data, and then select the model that gives the best fit to the entire dataset – although no (available) model may be entirely satisfactory. This simple approach tends to be conservative; that is, it tends to produce somewhat lower richness estimates.

As the above considerations show, the choice of model and tuning parameter interact, in a manner that does not currently have a direct statistical resolution. For a fixed value of τ, we can select a model using standard statistical criteria such as the Akaike Information Criterion (AIC). Given a variety of models, each fitted at several possible values of τ, one can examine the Pearson chi-square goodness-of-fit statistic, but at present the final choice of analysis must to some extent rely on expert judgment, pending further research. Fortunately, expansion of the set of available models (based on mathematical research) generally allows τ to increase, as the models become more flexible (Bunge and Barger 2008).

In summary, the advantages of the parametric stochastic abundance model approach are ease of interpretation, clear visual representation, and often excellent fit to a large proportion of the frequency count data. The disadvantages are: empirical selection of tuning parameter and model.

2.2 *Nonparametric Abundance Models*

In this approach, the basic structure is the same as above: each species is assumed to independently contribute a Poisson-distributed number of representatives to the sample, and the species' sampling intensities are assumed to follow some underlying "abundance" or "latent" distribution. However, here this distribution is not restricted to be a member of a parametric family (such as the lognormal), but rather is allowed to range over the entire set of possible distributions. Rather than fitting a parametric maximum likelihood estimate of total richness, this method computes a nonparametric maximum likelihood estimate or NPMLE (Wang and Lindsay 2005). There are several differences from the parametric approach.

- The final fitted abundance (sampling intensity) distribution is constructed by the procedure for each dataset, rather than being selected from a parametric family. (In fact the nonparametrically fitted distribution is discrete, essentially defining a small number of categories of species abundances.) This has the advantage of great flexibility, but gives no indication as to the suitability of well-known parametric models; essentially it builds a model de novo for each analysis.
- Since the method selects, or rather constructs an approximation to, a single model among the entire class of possible models (not just within a specific parametric family), the question of model selection among different parametric families does not arise.
- The NPMLE is relatively insensitive to the truncation point τ, so the question of τ, and its problematic interaction with model selection, does not arise, or is less important.
- There is not (at present) a direct variance (SE) formula for the NPMLE; instead the SE must be computed by some form of resampling such as the bootstrap. However, the bootstrap must be carried out very carefully, because the species richness estimation problem exhibits certain pathologies which may render bootstrap variance estimates inaccurate.
- There are several possible NPMLE's. All are based on the Poisson sampling model; the differences arise because of the various ways of stabilizing (reducing the variance of) the estimators, and in the approximations and algorithms used for computing them. Because the NPMLE approach is fairly recent, not enough experience has accumulated to make a definitive selection among these.
- It has been argued mathematically in a certain nonparametric theoretical framework that only lower bounds for the total richness can be regarded as reasonable (Mao and Lindsay 2007). Intuitively, this is because the nonparametric approach attempts to make minimal assumptions about the population structure (the underlying abundance distribution), and consequently one must always allow for

an arbitrary number of arbitrarily small species. Some versions of the NPMLE incorporate a "penalty parameter" or other devices to reduce this possibility.

As of this writing software to compute the NPMLE (and its variants) is not readily accessible, and therefore we do not compute the estimate for our example data. However, we expect software for this approach to enter the mainstream in the near future.

2.3 Coverage-Based Estimation

The *coverage* of the sample is the proportion of the population represented by the species that have been observed. For example, suppose there are four species in the population, say A, B, C, D, with proportional abundances 75%, 20%, 4% and 1%, respectively. If species A and C (only) appear in the sample then the coverage is 79%. (There is a statistical subtlety here: the *actual* coverage is random and depends on the particular sample observed, while the *expected* coverage is the long-term average coverage produced by the sampling procedure. Some methods address one version, some the other.) Estimation of coverage may be an easier statistical problem than richness estimation, and has its own literature (Mao 2004). The canonical example of a richness estimator based on this concept is the Abundance-based Coverage Estimator or ACE (Chao 2005). This estimator first inflates the observed number of species in inverse proportion to a nonparametric coverage estimate (the smaller the coverage, the greater the inflation), and then further adjusts nonparametrically according to the variability of the observed frequency counts. Many variations of ACE have been studied, including ACE1 for (apparently) higher-diversity populations (Chao and Lee 1992).

ACE and its variants:

- Are nonparametric and hence do not require model selection
- Admit direct variance estimation (SE), without resampling
- Are sensitive to the choice of τ – typically both the richness estimate and its SE increase with τ
- Provide a sequence of related estimators, typically assuming higher degrees of diversity in the population
- Are known to underestimate total richness in high-diversity populations (i.e., are downwardly biased in such cases)
- Can be theoretically (mathematically) related both to certain families of parametric models (such as the gamma-mixed Poisson/negative binomial), and to the NPMLE framework (Chao and Bunge 2002; Mao and Lindsay 2007)

In our example above, the preferred coverage-based estimator at $\tau = 10$ is ACE, which gives an estimate of 86 with SE of 14, and at $\tau = 16$ (the value selected by the parametric procedure) the preferred estimator is ACE1, which gives 152 with SE of 50. In some datasets the possible values of τ may vary by one or more orders of magnitude, with corresponding variation in the coverage-based estimates.

2.4 Discussion

Given the variety of approaches to estimation of species richness, which method should we use in a given case, and how should we select a final analysis to report? There are several principal considerations. First, the method used should correctly represent the researcher's operative assumptions about the underlying (target) population, sampling procedure, and biochemical system for production of the final frequency count data. All of the methods described above share the same fundamental framework: they assume that each species has a given sampling intensity, the intensities vary in such a way that they can be described by a probability distribution, and members of each species enter the sample independently according to their sampling intensities. If the data can be assumed to have been generated according to this framework, then all of the methods above are equally applicable. (We consider different data structures in Sect. 3).

The simplest and (in our view) best approach would be to compute all estimates and associated statistics such as SE's, goodness-of-fit assessments (where applicable), graphical representations, etc., and examine them all, because each method will illuminate a different aspect of the data. In the best-case scenario all methods will agree (approximately), leading to good confidence in the results. If there is strong divergence between the various analyses, this too is informative, and indicates that a more conservative approach should be preferred until further (or auxiliary) data becomes available. However, as of this writing there is no comprehensive and easy-to-use software package to carry out such a multimethod, parallel analysis. The coverage-based methods do not present major programming challenges and have been widely implemented in software. The parametric and nonparametric maximum likelihood methods require nontrivial algorithm and program development, and in some cases entail significant computing time, especially when many analyses are requested simultaneously, e.g., for multiple values of τ. At present, the existing software for these methods is not comprehensive or integrated, and the applied user will still need expert guidance, which may be available from the software authors. However, software for all methods is undergoing rapid development, and some software with reasonable user interfaces is beginning to appear, even for the computationally-intensive methods.

We recommend that the applied researcher make every effort *not* to be limited by easy-to-use, readily available software. Instead, we recommend that the researcher seek expert advice, which is available in most institutions from a statistics department or group, and analyze his or her data using as many of the above methods as is feasible, and in particular to compare the results of more than one parametric model (for example, the gamma-mixed Poisson or negative binomial model is usually too inflexible to accommodate the high-diversity data often encountered in microbial diversity research). The combined judgment of the biological and statistical experts can then be brought to bear on the results of the analyses. Such an approach will yield a range of results under slightly different model assumptions, and will tend to guard against over-optimistic acceptance of any single result, which may be biased downward or upward, unbeknown to the researcher.

Table 1 Comparison of abundance-based methods

Method	Advantages	Disadvantages
Parametric maximum likelihood	Well-understood properties; represents data via smooth distribution; responds stably to τ; natural visualization; tests suitability of specific abundance distributions	Model selection not obvious; results depend on (and model selection interacts with) τ; computationally intensive; patchy implementation in existing software
Nonparametric maximum likelihood	Does not require model selection; robust across a wide range of abundance distributions; apparently insensitive to τ	Procedure not yet thoroughly studied; standard error must be obtained by resampling; computationally intensive, and software not yet readily available
Nonparametric coverage-based	Well-understood properties; does not require model selection; robust across a range of abundance distributions; several user-friendly software implementations	Tends to be biased downward in high-diversity situations; sensitive to τ; little diagnostic information for choice of τ and specific estimator; no graphical representation

Table 1 summarizes the salient pros and cons of the various methods at the present time.

We emphasize that while this is the current state of the art, progress is rapid and we expect many of the disadvantages listed above (for all methods) to be ameliorated in the near future, especially in terms of computation.

3 Incidence Data

In this scenario we collect species occurrence or incidence data on several different sampling "occasions," or from several different lists; this is also known as capture–recapture, multiple recapture, or multiple list data. For example, the Table 2 shows 10 different samples or lists, which yielded a combined total of 15 observed species. (This is a subset of a larger dataset with 46 samples and 3,717 observed species, extracted from GenBank at the 90% similarity level (Epstein and Bunge 2008).) Each row represents the "capture history" of a particular species (arbitrary species ID numbers are assigned), and each column represents the list of species observed on a given sampling occasion. Note that on each occasion, the only presence or absence – the "incidence" – of each species is recorded, where 1 indicates that the given species was observed on the given occasion (0 otherwise). The right-most column gives the total number of observations for each (observed) species. Analysis of such data has a long history and immense literature, dating back at least as far as the eighteenth century (Borchers et al. 2002, Chao and Huggins 2005); here we attempt only a sketch from a particular point of view.

Table 2 Example multiple recapture data

	Sample										
Species ID	1	2	3	4	5	6	7	8	9	10	Total
1	0	1	0	0	0	0	0	0	1	0	2
2	0	0	0	0	0	0	0	0	1	0	1
3	1	1	1	0	0	0	0	0	1	0	4
4	0	0	1	0	0	0	0	0	0	0	1
5	1	0	0	0	0	0	1	1	1	0	4
6	0	0	0	0	0	1	0	0	0	0	1
7	1	1	1	0	0	1	0	0	1	0	5
8	0	0	1	0	0	0	0	0	1	0	2
9	0	0	0	0	0	1	0	0	1	0	2
10	0	0	0	0	0	1	0	0	0	0	1
11	0	1	0	0	0	0	1	0	1	0	3
12	0	0	0	0	0	0	0	0	0	1	1
13	0	0	0	0	0	0	0	0	1	0	1
14	0	0	0	0	0	0	0	0	1	0	1
15	0	0	0	0	0	0	0	1	0	0	1

We note that in some cases the actual "abundance" or number of observations of a given species on a given occasion may be recorded, leading to integer entries greater than 1; this may be called "multiple abundance data," although there is no standard terminology in this case. Clearly multiple incidence data can be derived from multiple abundance data, but not the reverse. Note also that frequency-count data can be derived from the marginal totals, but the full table cannot be recovered from the frequency-count data. In this example, the latter is (1,8), (2,3), (3,1), (4,2), (5,1). Here the maximum possible frequency is equal to the number of samples. It is possible to apply frequency-count methods to such data if the number of samples is large enough; we return to this issue below.

Multiple incidence data is more highly structured than frequency-count data, and its statistical analysis admits more variations; here we only attempt an outline of the areas we see as especially relevant to microbial ecology. Much of the literature in this area originated with population size estimation for macro-fauna – birds, mammals, fish – and in this application a row in the table represents the capture history for a particular animal. A certain taxonomy of models has emerged from this literature (Borchers et al. 2002). It is not ideally adapted to microbial ecology applications but it has become a de facto standard, and (at least) the first four models are noteworthy here.

- M_0: Global homogeneity. Each species is equally likely to occur, on each occasion, i.e., each species has the same sampling intensity or abundance, and the "sampling effort" is the same on each occasion.
- M_t: "Time" (t) effect. All species have the same sampling intensity or abundance, but sampling effort varies with occasion or time. Thus all species have the same probability of occurrence on a given occasion (within a given column of the table), but this probability varies across occasions (columns).
- M_h: Heterogeneity (h) effect. The species have different (heterogeneous) sampling intensities or abundances, but sampling effort is the same on each occasion.

Thus a given species' probability of occurrence is the same on every sampling occasion: the probability of occurrence is the same within a given row of the table, but varies from row to row.

- M_{th}: Time and heterogeneity effects. The species have different sampling intensities, *and* sampling effort varies across occasions. Thus the occurrence probabilities vary both across rows and across columns.

Further models involving behavioral (*b*) effects have been studied in the literature. These were originally intended to account for individual animals' responses to being captured (e.g., becoming "trap-happy" or "trap-shy"), and while the statistical models for such effects may have an alternative interpretation in the microbial ecology setting, we do not discuss them here. In the following discussion we assume that the sampling occasions or lists are (statistically) independent. There are many situations in which this assumption may be false, but we will retain it here for simplicity, and because it is often reasonable in microbial ecology.

The models M_0 and M_t, which assume equal species abundances, admit straightforward maximum likelihood estimates and can be dealt with in a reasonably uncomplicated manner, statistically and computationally. Unfortunately, the equal-abundance assumption is rarely if ever realistic, and causes severe downward bias when the true population is heterogeneous. Thus the models of interest here are M_h and M_{th}. The more general of these is of course M_{th}; however, it may possible for the researcher to specify, on substantive grounds, whether the "time" effect should be assumed to be present or not, that is, whether sampling effort can be assumed to be constant across occasions. Thus both M_h and M_{th} are potentially useful.

The same three classes of estimators – parametric maximum likelihood, nonparametric coverage-based, and nonparametric maximum likelihood – exist for incidence data as well as abundance data. If we restrict our attention to models M_h and M_{th}, this gives six potential families of statistical procedures to consider. We will briefly consider each of the six possibilities. In addition, there is (at least one) alternative approach based on "estimating equations," which has both parametric and nonparametric aspects, and admits an elegant extension even to the most general model M_{bth} (Chao et al. 2001). However, in its present form this method depends on the "time" order of the samples, i.e., it is not invariant to permutation of the lists. Since this assumption appears to be more adapted to certain kinds of animal-trapping surveys, we do not delve into it in detail here.

3.1 Parametric Incidence Models

3.1.1 Model M_h

In this case, the row totals of the data, that is, the number of times each species is observed, are binomial random variables, where the binomial "success probability" (the probability that a given species is observed on a given occasion) varies from row to row. A parametric mixture model, analogous (and in some ways equivalent) to the mixture models for frequency data discussed above, asserts that the success probabilities vary according to some "mixing" (probability) distribution. This gives rise to a "mixed binomial" model. While there is an unlimited spectrum of possible mixing distributions, only a few are commonly used at present. The most common is the beta-mixed binomial or beta-binomial. This distribution is mathematically tractable, and its computation is often feasible (though nontrivial), but it does not seem to have a convincing justification otherwise. In addition, it may fail to fit a given dataset, or readily available computational algorithms may fail to converge, yielding no results. A second family of candidates consists of finite mixtures (Pledger 2005), which essentially model the population using a small number of "abundance classes": for example, a two-component finite mixture postulates the existence of a group of low-abundance species and a group of high-abundance species (approximately). These models appear to show promise in terms of empirical applicability even when only two or three components are used. In the example above, the beta-binomial analysis yields an estimate of 25 species, and the two-component mixture yields 26 (both with SE $\geq$ 3), with the latter fitting much better than the former.

3.1.2 Model M_{th}

Here the success probability varies both with species and with sampling occasion, i.e., both with row and column. In order for species richness estimation to be feasible some structure must be imposed. Again there is an infinite spectrum of possible models, but the "Rasch" model is at least conceptually simple, potentially applicable, and reasonably well-studied in a variety of contexts (Fienberg et al. 1999). Let p_{ij} denote the probability that species i appears on occasion j; then one simple version of the Rasch model asserts that $\log(p_{ij}/(1-p_{ij})) = \alpha_i + \beta_j$, so that α_i is the species heterogeneity (row) effect, and β_j is the sampling occasion (column) effect. Further structure is still required for richness estimation, and it can be imposed in various ways. Often the effects α_i and β_j are taken to be random variables with some parametric distribution (in terms of the α_i, this again leads to a mixture model for the species abundances); the model can then be fitted by maximum likelihood, or by various levels of hierarchical Bayesian procedures, which are beyond our scope here. However, at present there is no readily available, easy-to-use software for any version of the Rasch model for species-richness estimation. Existing software is specific to particular model assumptions, and requires a statistical expert for use and interpretation.

3.2 *Nonparametric Incidence Models*

For model M_h, typically a mixing distribution is assumed for the species' detection probabilities, as described above. It is possible to specify this distribution nonparametrically. That is, rather than estimating the parameters of a fixed family of distributions such as the beta-binomial, these statistical procedures attempt to approximate the mixing distribution without making any particular restrictions on its form. Such a nonparametric maximum likelihood procedure was first put forward for this model by Norris and Pollock (1996) (see also Tardella 2002 and Huggins and Yip 2001), and in principle shares the advantages of the NPMLEs discussed above, but again software has not yet become generally available. We have not found an NPMLE method for model M_{th} in the literature to date.

3.3 *Nonparametric Coverage-Based Incidence Methods*

As with abundance data, these procedures inflate the observed number of species by the reciprocal of a nonparametric estimate of the sample's coverage, and then adapt nonparametrically to the apparent heterogeneity in the data (assessed via an estimate of the coefficient of variation). Again, the methods are primarily due to Chao and co-authors (Lee and Chao 1994). The procedure for model M_h is known as ICE, or Incidence-based Coverage Estimator, and it also has a high-diversity variant ICE1; there also exist moderate- and high-diversity procedures for model M_{th} although these do not have specific conventional acronyms. As was true for abundance data, these methods require a user-selected maximum frequency τ, and the results are sensitive to τ, although little guidance is available for its choice apart from selecting the default value of $\tau = 10$. These methods are (computationally) relatively straightforward and have been implemented in several readily-available and usable software packages. For our example above (which is not highly heterogeneous), ICE gives an estimate of 24 with SE 3, and the M_{th}-based estimate is 25 with SE 8. For this example the various procedures agree reasonably well, but as with abundance data, careful comparison of models and estimators is generally required.

3.4 *Models with Covariates*

There is a substantial amount of research on statistical models for incidence data that can accommodate covariates, i.e., "external" information about the nature of the various sampling occasions or situations, or even about some or all of the observed species themselves. For example, some measure of (or proxy for) "sampling effort" might have been recorded on each occasion, and we might seek to incorporate this information into the analysis. Such methods vary widely and tend to be adapted to fairly particular situations, and hence we do not attempt to describe them here; see, e.g., Zwane and Van der Heijden (2005).

4 Remarks

Given the results of a statistical analysis, to what larger population do they apply? This question is especially pressing in ecological problems of species-richness or animal abundance estimation, because it can be difficult to precisely define the population from which the sample emanated. In the case of species richness estimation in microbial ecology, there are often two levels of sampling: first, a sample of material is taken from some natural environment, say, 5 L of ocean water from a bay. Then a subsample of water is taken from this sample, an rRNA gene clone library is generated from this subsample, species are identified using rRNA gene sequences, and finally species frequency count data is obtained. The statistical projections based on the frequency count data can only strictly be said to apply to the 5-L sample, not to the entire bay.

In this connection the concept of "homogeneous mixing" may be of some value (here the term "mixing" is used in the ordinary sense rather than the technical sense above). The mathematical model for generation of abundance data states that there are multiple simultaneous independent homogeneous Poisson processes, one for each species, and that the members of a given species appear in the sample according to the rate of the process; the sampling intensity is the expected number of members of that species expected to appear during one unit of sampling effort. From an applied statistical perspective, this means in particular that the results of a statistical inference based on frequency-count (abundance) data can be extended to a stationary, fixed population. That is, the sampling model is valid if the population does not change during the sampling period, and the population is "homogeneously mixed" – the organisms in question are uniformly dispersed throughout the sampled environment. This clarifies why, in the example above, the projections apply to the 5-L sample but not to the ocean environment from which the sample was drawn; the 5 L sample can be reasonably considered to be homogeneously mixed, but the surrounding ocean cannot.

Similar considerations arise in the case of incidence samples. In particular the models and procedures discussed herein are all for "closed" populations, which again are assumed to be fixed, without migration, birth or death, at least during the sampling period. Extensions to nonclosed populations exist; see Borchers et al. (2002) for an introduction.

A related issue is that of statistical independence within the sample. For abundance samples, we assume that the individuals of the various species enter the sample independently. At present little theoretical work has been done to weaken this assumption to (various forms of) dependence, and statistical tests for independence are not (yet) available in this problem. It is not known how sensitive the analyses are to violations of this assumption. For incidence sampling the picture is a little clearer. While most incidence models assume independence within columns, i.e., the species (or animals) enter the sample independently on a given occasion, and (as for abundance data) this assumption is difficult to assess, the models have been extended to admit dependence between columns. Such models originated in the animal population estimation literature, where an animal may become "trap-happy" or "trap-shy" resulting in dependence between its behavior on successive sampling occasions. Hence the taxonomy has been extended to allow a subscript *b* for behavior, with correspondingly more complex models M_{bt}, M_{bh}, and so on. However, in microbial ecology the assumption of independent columns (samples) is often plausible, so we do not enter into the dependent models here; see, e.g., Chao and Huggins (2005).

Finally, suppose that we have arrived at a provisionally satisfactory probabilistic model to describe the generation of our data (at least approximately). As discussed above, for a given model there are several valid statistical procedures that may be used to analyze the data and make inference about the population, i.e., about the true species richness. Which should the researcher choose? We recommend that several competing procedures be computed, and the results be compared. This is because the process of interpreting the differences between the analyses will illuminate the data (and the particular problem at hand) from new angles. Indeed it is in the comparison of results, taking into account the differing assumptions and structural properties of the statistical procedures, that the most valuable insights may in some cases be gained. Ultimately a single estimate of species richness must be reported, along with a corresponding standard error and (possibly) confidence interval. The point is, however, that this choice should be made after evaluating the competitors in view of their advantages and disadvantages, rather than a priori for reasons of simplicity, ease of use of software, etc. As the theoretical study of these methods progresses, and software becomes more readily available, such a comparative approach will become more accessible to the applied researcher.

APPENDIX: Software

Here is a list of some computer software that is available as of this writing. Documentation is available at the websites listed. The collection of such software continues to expand, and for new applications and data analyses the reader should check recent developments via expert advice and Internet searching.

- Abundance data
  - Code written in MAPLE (www.maplesoft.com) for various parametric models, http://www.stat.cornell.edu/~bunge/(Hong et al. 2006)
- Abundance or incidence data
  - SPADE, http://chao.stat.nthu.edu.tw/softwareCE.html(Shen et al. 2003)
  - EstimateS, http://viceroy.eeb.uconn.edu/EstimateS(Colwell 2005)
- Incidence data
  - DENSITY, http://www.landcareresearch.co.nz/(Efford et al. 2004)
  - CARE-2, http://chao.stat.nthu.edu.tw/softwareCE.html(Chao et al. 2001)
  - CAPTURE, http://www.mbr-pwrc.usgs.gov/software.html(Rexstad and Burnham 1991)
  - M-SURGE, http://www.cefe.cnrs.fr/(Choquet et al. 2004)

References

Behnke A, Bunge J, Barger KJ, Stoeck T (2008) Impact of the time dimension on our perception of microbial molecular diversity and its patterns. Submitted for publication

Borchers DL, Buckland ST, Zucchini W (2002) Estimating animal abundance: closed populations. Springer New York

Bunge J, Barger K (2008) Parametric models for estimating the number of classes. *Biometrical Journal* 50 (5)

Chao A (2005) Species estimation and applications. In: Balakrishnan N, Read CB, Vidakovic C (eds) Encyclopedia of statistical sciences, 2nd edn, vol 12. Wiley, New York, pp 7907–7916

Chao A, Bunge JA (2002) Estimating the number of species in a stochastic abundance model. *Biometrics* 58:531–539

Chao A, Huggins RM (2005) Classical closed population models. In: Manly B, Mcdonald T, Amstrup S (eds) The handbook of capture–recapture methods, Princeton University Press, Princeton, pp 22–35

Chao A, Lee S-M (1992) Estimating the number of classes via sample coverage. *J Am Statist Assn* 87:210–217

Chao A, Yip PSF, Lee S-M, Chu W (2001) Population size estimation based on estimating functions for closed capture–recapture models. *J Statist Plan Inference* 92:213–232

Choquet R, Reboulet A-M, Pradel R, Gimenez O Lebreton J-D (2004) M-SURGE: new software specifically designed for multistate capture–recapture models. Anim Biodivers Conserv 27:207–215

Colwell RK (2005) EstimateS: Statistical estimation of species richness and shared species from samples. Version 7.5. er's Guide and application published at: http://purl.oclc.org/estimates

Efford MG, Dawson DK, Robbins CS (2004) DENSITY: Software for analysing capture–recapture data from passive detector arrays. *Anim Biodivers Conserv* 27:217–228

Epstein SS, Bunge J (2006) Estimation of microbial diversity from GenBank data. *Appl Environ Microbiol* 72(10):6578–6583

Epstein SS, Bunge J (2008) Estimation of microbial diversity from GenBank data. In preparation.

Fienberg SE, Johnson MS, Junker BW (1999) Classical multilevel and Bayesian approaches to population size estimation using multiple lists. *J R Stat Soc: Ser A* 162:383–405

Hong S-H, Bunge J, Jeon S-O, Epstein SS (2006) Predicting microbial species richness. *Proc Natl Acad Sci USA* 103:117–122

Huber JA, Mark Welch DB, Morrison HG, Huse SM, Neal PR, Butterfield DA, Sogin ML (2007) Microbial population structures in the deep marine biosphere. *Science* 318:97–100

Huggins RM, Yip PSF (2001) A note on nonparametric inference for capture–recapture experiments with heterogeneous capture probabilities. *Statistica Sinica* 11:843–853

Lee S-M, Chao A (1994) Estimating population size via sample coverage for closed capture–recapture models. *Biometrics* 50:88–97

Magurran AE (2004) Measuring biological diversity. Blackwell, Oxford

Mao CX (2004) Predicting the conditional probability of discovering a new class. *J Am Stat Assoc* 99:1108–1118

Mao CX, Lindsay BG (2007) Estimating the number of classes. *Ann Stat* 35:917–930

Norris JL III, Pollock KH (1996) Nonparametric MLE under two closed capture–recapture models with heterogeneity. *Biometrics* 52:639–649

Pledger S (2005) The performance of mixture models in heterogeneous closed population capture–recapture. *Biometrics* 61:868–876

Rexstad E, Burnham KP (1991). User's Guide for Interactive Program CAPTURE. Colorado Cooperative Fish and Wildlife Research Unit, Fort Collins, ***CO, USA, pp 29

Shen TJ, Chao A, Lin CF (2003) Predicting the number of new species in further taxonomic sampling. *Ecology* 84:798–804

Stackebrandt E, Goebel BM (1994) Taxonomic note: A place for DNA:DNA reassociation and 16S rRNA sequence analysis in the present species definition in bacteria. *Int J Syst Bacteriol* 44:846–849
Tardella L (2002) A new Bayesian method for nonparametric capture–recapture models in presence of heterogeneity. *Biometrika* 89:807–817
Wang J-PZ, Lindsay BG (2005) A penalized nonparametric maximum likelihood approach to species richness estimation. *J Am Stat Assoc* 100:942–959
Williamson M, Gaston KJ (2005) The lognormal distribution is not an appropriate null hypothesis for the species-abundance distribution. *J Anim Ecol* 74:409–422
Zwane E, van der Heijden P (2005) Population estimation using the multiple system estimator in the presence of continuous covariates. *Stat Modelling* 5:39–52

Characterizing Microbial Population Structures through Massively Parallel Sequencing

Mitchell L. Sogin

Contents

Abstract Rank abundance plots of microbial populations are long-tailed distributions in which a few highly abundant populations account for most of the organisms in a community, whereas lower-abundance taxa represent a large number of different phylotypes. New massively parallel DNA pyrosequencing techniques that routinely sample many hundreds of thousands of homologous genes permit the detection of lower abundance taxa even in the presence of overwhelming levels of dominant microbial populations. The technology provides estimates of both richness (diversity) and evenness (relative abundance of different kinds of microbes). When applied to studies of marine and soil microbial populations, the massively parallel sequencing of many thousands of rapidly evolving regions in ribosomal RNA genes reveals that microbial diversity is 1–2 orders greater than all previously published estimates from traditional capillary DNA sequencing methodologies. Most of the diversity is accounted for by low-abundance phylotypes that constitute the rare biosphere. These taxa might represent "seed" organisms that under shifts in ecological conditions can become more abundant, or they might be persistently rare over large temporal and spatial scales.

M.L. Sogin
Josephine Bay Paul Center, Marine Biological Laboratory at Woods Hole, 7 MBL Street, Woods Hole MA 02543, USA
e-mail: sogin@mb.edu

Microbiol Monogr, doi:10.1007/7171_2008_12

1 Microbial Diversity and Habitability

Microbes are the most abundant cellular life forms on Earth. Their cell numbers in terrestrial and aquatic environments exceed 5×10^{30} organisms (Whitman et al. 1998), which eclipses the total number of plant and animal cells by at least 2–3 orders of magnitude. They live inside of us, inhabit our food and water, disperse in our air, and grow on surfaces that we contact every day. Microbes are mostly innocuous and many are essential for our very existence. Without bacteria, archaea, and protists, multicellular life on earth would not have evolved and would not be sustainable. From the time of their origins, microscopic factories – initially anaerobic and later aerobic – have served as essential catalysts for all of the chemical reactions within biogeochemical cycles that shape planetary change and habitability. Microbial carbon remineralization, with and without oxygen, maintains the carbon cycle. Microorganisms control global utilization of nitrogen through nitrogen fixation, nitrification, and nitrate reduction. They drive the bulk of sulfur, iron, and manganese biogeochemical cycles (Atlas and Bartha 1993) and regulate the composition of the atmosphere, influence climates, impact water quality, recycle nutrients, and decompose pollutants. They exert a significant economic impact in agriculture, fisheries, animal husbandry, a vast array of fermentation processes, and in the biotechnology industry including the production of pharmaceuticals. Yet some microbes in the environment have the potential to cause disease in humans and in macrobiota.

2 The Molecular Influence on Microbial Taxonomy and Ecology

We are just beginning to discover what promises to be an extraordinarily vast number of heretofore-unknown microbes and microbial activities. Microbial ecologists seek to understand "Who is there? What are they doing? Where did they come from? How do microbial populations shape earth's habitability?" Answers to these questions require detailed descriptions of microbial population structures that include measures of richness (diversity) and evenness (relative abundance of different kinds of microbes in a community). During the first 100 years of microbiology, the lack of agreement about which phenotypic features were most valuable for defining microbial systematics led to conflicting views of their natural relationships and gross underestimates of diversity that encompassed a mere 5,000 bacterial "species" (Guerrero 2001). Thirty-five years ago, Woese pioneered the use of ribosomal RNA (rRNA) sequence comparisons for charting the evolutionary history of microorganisms (Sogin et al. 1971; Woese and Fox 1977; Woese et al. 1974). The prokaryote–eukaryote dichotomy soon gave way to the three domain system, which includes Bacteria, Archaea, and Eukarya (Woese et al. 1990). With the introduction of improved sequencing technology, Pace developed culture-independent molecular tools for assessing the diversity and ecology of microorganisms (Pace 1997; Pace et al. 1985). He demonstrated that the sequence of an rRNA gene isolated from environmental DNA extracts could serve as proxies for the presence of a microbe in a sample. Several technical advances accelerated progress. The first was the demonstration by Medlin et al. (Medlin et al. 1988) that polymerase chain reaction (PCR) technology could efficiently amplify ribosomal rRNA genes from the primary domains and thereby eliminate the requirement to screen recombinant genome libraries. The second was the development of automated DNA sequencing technology (Smith et al. 1986). The third was the introduction of powerful tools for aligning DNA sequences and inferring phylogenetic frameworks (Felsenstein 1989; Higgins 1994; Swofford 2001).

Sequence analyses of clones containing rRNA sequences from shot-gun genome libraries or PCR amplicon libraries of environmental rRNA genes have become the "gold standard" for assessing species richness in microbial communities. Current databases contain more than 500,000 rRNA sequences that correspond to phylotypes (Table 1) from diverse microbes (http://www.arb-silva.de,, http://bioinformatics.psb.ugent.be/webtools/rRNA/,http://www.ncbi.nlm.nih.gov/http://www.rdp.cme.msu.edu). The number of bacterial phyla has increased from Woese's description of a dozen lineages to more than 100 major phyla (Fig. 1) – most of which do not include cultured representatives (Ley et al. 2006). On the basis of surveys of environmental rRNA gene sequences, microbial diversity is at least 100–1,000 times greater than estimates based upon cultivation-dependent surveys (Pace 1997). Yet, the molecular assays have only captured a fraction of microbial diversity and they rarely provide estimates of relative abundance for different kinds of microbes or operational taxonomic units (OTUs) (Pedros-Alio 2006).

The expense of sequencing full-length rRNA genes has constrained the size of most molecular surveys. Determining a full-length rRNA gene sequence typically requires

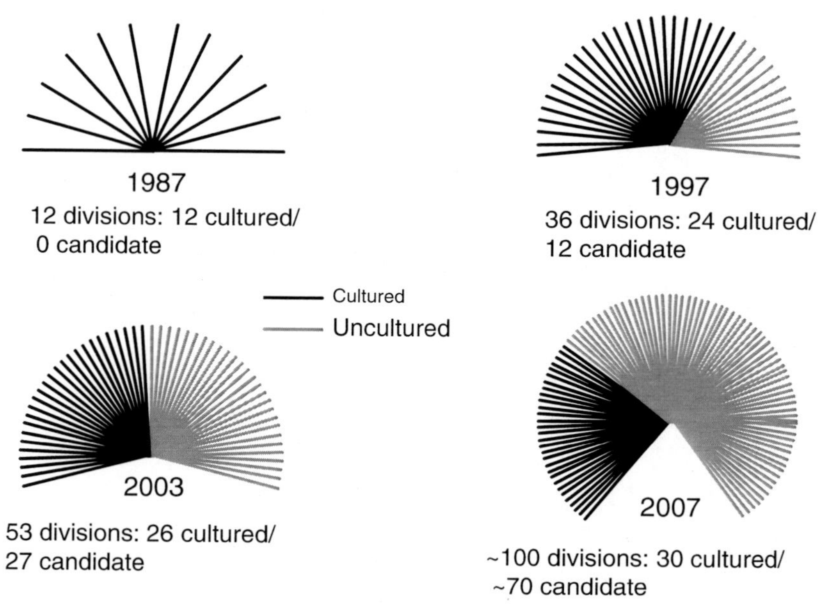

Fig. 1 Bacterial rRNA diversity from cultivars and environmental DNA samples

4–8 reads on an automated platform. Early automated DNA sequencing systems employed slab electrophoresis that generated fewer than 100 sequences per day at a cost of $5–30 per read. The introduction of capillary-array electrophoresis (CAE) sequencing increased read accuracy and reduced costs to less than $1 per read on Molecular Dynamics or ABI platforms that could process >1,000 templates per day. The promise of discovering new phylotypes fostered experimental designs where low-resolution procedures, e.g., restriction fragment length polymorphisms, identify putatively distinct clones for DNA sequencing. This binning procedure allows for more cost-efficient surveys of microbial diversity (since high frequency clones with identical sequences are excluded, fewer clone sequences are required to discover novel taxa), but it comes at the expense of minimizing information about the relative abundance of distinct phylotypes. Despite the increased efficiency of CAE, few molecular microbial diversity surveys collect more than 1,000 full-length sequences. This modest amount of data can detect only a small fraction of the 10^8–10^9 microbes per liter commonly found in aquatic environments or the 10^9 microbial cells in a gram of soil (less than one per million cells). Such studies detect sequences from high-abundance organisms that likely represent rapidly growing populations. They fail to report sequences from most of the low-abundance taxa, and some of these may be the most interesting for predicting shifts in microbial population structures in response to ecological change. The occurrence and distribution of the low-abundance taxa remain undersampled and uncharted (Pedros-Alio 2006).

The undersampling problem is even more severe for metagenomic studies, where the sequencing effort is distributed across the entire genome rather than targeted to a single gene. Metagenomic investigations usually seek to characterize the metabolic properties of microbial communities that define ecosystem function. These surveys commonly collect 100–200 MB (mega base pairs) of sequence data or approximately 100× coverage for the average 2 MB microbial genome. A threefold coverage of any particular genome will statistically describe 90% of its functional genes, but knowing the genomic context of any annotated gene will require 8–15-fold genome coverage in order to assemble reasonable size contigs. In a typical 100–200 MB genome study, descriptions of metabolic diversity will be successful only for microbes that account for 5% of the genomes in a sample and assembly will be possible only for genomes that represent at least 15% of the genomes in a sample. This agrees with the general observation that despite deep sampling of sequence space, the rate of new protein-family discovery in metagenomic surveys remains linear (Yooseph et al. 2007). Yet microbial communities are far more complex with richness estimates in the many thousands of taxa.

3 Massively Parallel DNA Pyrosequencing and Microbial Diversity

As an alternative to molecular analyses of cloned PCR amplicons, several "Next generation" DNA sequencing technologies (Illumina's Solexa, Applied Biosystems' SOLiD, and 454 Pyrosequencing technology implemented on Roche's Genome Sequencers) reduce the costs of DNA sequencing by 2 or more orders of magnitude through the use of massively parallel strategies that circumvent the requirement to clone DNA templates. Despite their short 25–35 base pair (bp) read lengths, the enormous capacity (60–200 million reads/run for Solexa and SOLiD systems, respectively) allows for cost-effective sequencing of microbial genomes from cultivars through the assembly of large numbers of overlapping reads (Butler et al. 2008). However, for analyses of environmental DNAs, 25–35 bp read lengths are not sufficient for identifying or annotating homologous coding regions. Since each sequence read can theoretically represent DNA from a distinct gene from a complex mixture of microbial genes, their assembly into longer contigs risks the formation of sequence chimeras. As a compromise, the massively parallel capability of 454 Life Science's pyrosequencing technology (Margulies et al. 2005) implemented on the Roche Genome Sequencer (GS) can generate 300,000–450,000 sequences in a single run. The original GS20 produced sequences that contained ~100 bp, but the read length has increased on the GSFLXto 230 bp with expectations of achieving one million 400 bp reads/run in the near future.

For the analyses of PCR amplicon libraries from environmental DNA samples, the primer design must include the 454 Life Science A primer fused to the conserved 5′ region and a reverse 3′ primer that flanks the targeted genomic sequence. The Life Science A primer sequence on each fragment in the amplicon library anneals to complementary oligonucleotides that are tethered onto micron-size beads. The annealing conditions favor one fragment per bead. The beads are emulsified in a PCR mixture in oil, and PCR amplification (emPCR) occurs in a microreactor generating approximately ten million copies of a unique DNA template. After breaking the emulsion, the DNA strands are denatured and beads carrying single-stranded DNA clones are deposited into wells on one of the two regions of the PicoTiterPlate for pyrosequencing on a Roche Genome Sequencer. Each PicoTiterPlate contains 1.6 million wells that can hold a bead and additional reagents including polymerase, luciferase, and ATP sulfurylase. Microfluidics cycle each of the four nucleotide triphosphates over the PicoTiterPlate, and incorporation of a nucleotide releases pyrophosphate, the substrate for a luminescence reaction (Ronaghi et al. 1998). A cooled charge-coupled device (CCD) camera records the luminescence reaction that produces a signal proportionate to the amount of released pyrophosphate. The record of intensity of each flow of a nucleotide is a flowgram, analogous to a chromatogram, that reports the order of A, C, G, and T residues from a DNA sequencing template. Flowgram values correspond to the homopolymer length for that base. On average, 300,000–450,000 of the wells produce usable reads.

This new methodology brings with it different sources of error than traditional CAE sequencing. For example, it is sometimes difficult to resolve the intensity of luminescence produced when a homopolymer is encountered. The result can be ambiguity of homopolymer length, particularly for longer homopolymers. In addition, insufficient flushing between flows can cause single base insertions (carry forward events) usually near but not adjacent to homopolymers. Finally, low nucleotide concentrations within a flow can cause incomplete extension within homopolymers. Margulies et al. (Margulies et al. 2005) found that an excess of intermediate flowgram values indicates a poor-quality read. For genome sequencing, the error rate for single reads can be as high as 5 or 6%, but the assembly of many overlapping reads can produce accurate consensus sequences (Goldberg et al. 2006; Moore et al. 2006) with error rates as low as 0.03%. However, the generation of consensus sequences to improve accuracy is not appropriate for studies that seek information about natural variation from every read. For example, in metagenomic (Edwards et al. 2006) or PCR amplicon (Sogin et al. 2006) libraries from environmental DNA samples, each sequence read can theoretically represent DNA from a distinct gene from a complex mixture of microbial genes. Using a model system that compared tags from a set of clones of known sequence, Huse et al. demonstrated an error rate (incorrect bases/total number of expected nucleotides) of 0.49% on the original GS20, and identified several factors that can be used to remove low-quality reads, improving the accuracy to 99.75% or better (Huse et al. 2007). By using objective criteria to eliminate low-quality data (elimination of reads with one or more N's or reads that were shorter than 50 nt or longer than 100 nt), the quality of individual GS20 sequence reads in molecular ecological applications can surpass the accuracy of traditional CAE methods.

The 454 pyrosequencing technology is particularly adept at describing microbial population structures through the analyses of short "tag" sequences (Sogin et al. 2006) from one or more rRNA hypervariable regions (Fig. 2). Each tag serves as a proxy for the occurrence of an rRNA coding region in an environmental DNA sample. Enumerating the frequency of individual tags will provide a first-order description of the relative abundance of different microbes in a community. Rarefaction of clusters of closely related sequence tag sequences and nonparametric estimators such as Chao1 (Chao 1984) and the abundance-based coverage estimator ACE provide estimates of species richness (Chao 1992; Chao et al. 1993). The highly variable nature of the tag sequences and paucity of positions do not allow direct inference of phylogenetic frameworks; however, when tag sequences serve as a query against a reference database of hypervariable regions from full-length rRNA sequences of known phylotypes, it is possible to extract information about taxonomic identity, microbial diversity, and relative abundance (Huber et al. 2007; Sogin et al. 2006). Matches of each tag sequence to hypervariable regions from known phylotypes (using a reference data base of ~500,000 entries that span a targeted hypervariable region) provide information about taxonomic identity. The frequency of individual tags provides an estimate of the relative abundance of specific microbes in a population. The reproducibility is excellent and PCR bias minimally impacts the differential amplification of these very short rRNA regions from complex microbial populations. The major constraint is the taxonomic resolution inferred for longer environmental rRNA sequences that currently dominate the molecular databases. In some cases, the taxonomic assignments of environmental sequences map to genus; others resolve only to class or phylum.

4 Bioinformatic Interpretation of Pyrosequences from rRNA Hypervirable Regions

The first application of this strategy (Sogin et al. 2006) targeted the rRNA V6 region (*E. coli* positions 967–1,046), which ranges in size from 60–80 bp. The V6 information content is analogous to the use of hypervariable regions in DNA microarray

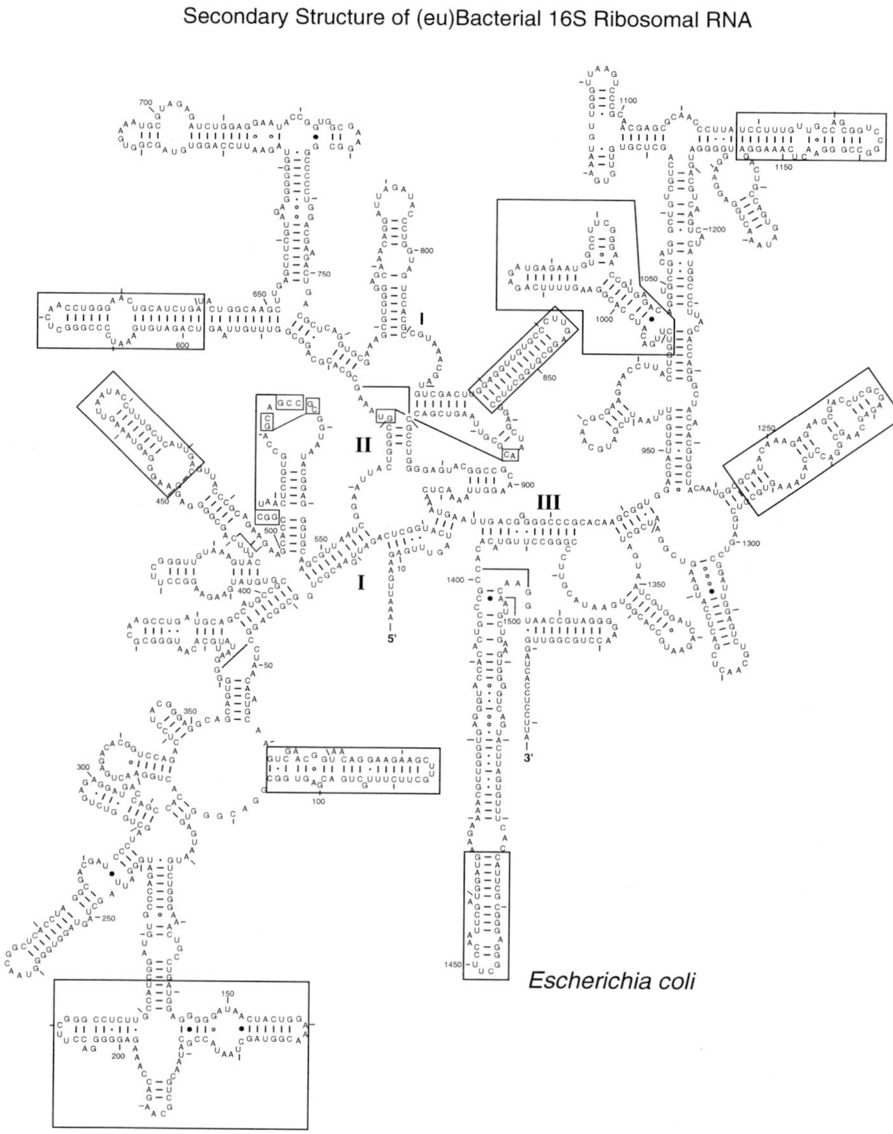

Fig. 2 V6 Region of SSU rRNA. Conserved sequences flank the hypervariable V6 regions of bacterial rRNAs and serve as primer sites to generate PCR amplicon

experiments that survey the presence of gene sequences from specific microbes (DeSantis et al. 2007; DeSantis et al. 2005). Both technologies use 60–80 bp DNA sequences to measure the presence of specific rRNA genes in a DNA sample. The principle difference is that the range of taxa detected by tag sequencing is not constrained by OTU-specific probe sequences in a DNA microarray. With the capacity to generate >240 bp reads on the GSFLX, it is now possible to target longer hypervariable regions (Gutell et al. 1985; Van de Peer et al. 1996), e.g., the ~130 bp V3 and V9, or a combination of hypervariable regions such as V2 plus V3. The reproducibility is excellent, and PCR bias minimally impacts the differential amplification of these very short rRNA regions from complex microbial populations (Huber et al., manuscript in preparation). Other attributes include (1) ~100-fold lower cost than traditional sequencing of rRNA amplicons (reagent costs ~$0.01 per GS-FLX read versus $1–2 per capillary read), (2) the ability to detect minor members in a microbial population, and (3) fewer CPU cycles to infer phylogeny; the analyses rely upon identification of best matches with rRNAs from precomputed phylogenies.

For most surveys, the collection of 25,000–50,000 tags will adequately describe microbial population structures for a microbial community. Multiplexing or "bar coding" strategies allow the concurrent collection of thousands of tags from multiple samples in a single 4-h sequencing run without use of partitioning gaskets that reduce the number of sequencing wells on the PicoTiterPlate. By incorporating a unique 5-bp "key" between the 454 Life Sciences A-primer and the 5′ flanking rRNA primer (Fig. 3), each PCR reaction produces products that contain a discrete 5-bp "key". After combining multiple PCR amplicon libraries for pyrosequencing, the 5-bp "key" identifies the sample source for each tag sequence. The use of a 5-bp "key" allows for the synthesis of as many as 81 oligonucleotides that differ at a minimum of two sites. To capture the full diversity of rRNA sequences represented in molecular databases, it is necessary to use a cocktail of five primers at the 5′ end and four primers at the 3′ end of the V6 region. The preparation of at least three independent PCR cocktails containing as little as 5 ng of DNA from each sample will minimize the impact of potential early-round PCR errors.

The bioinformatics analysis of tag sequences includes deconvolution of sequences according to bar coded primers, removal of low-quality sequences, assignments of taxonomic identities to tag sequences, and rarefaction analyses of assigned taxonomic units or sequence-based OTUs. Bioinformatics searches use the 5-nt keys to deconvolute the sequence reads into separate sample bins. Quality control routines

5′ Primers

|454 Primer A | | 5′ rRNA Primer |

gcctccctcgcgccatcag NNNN NCNACGCGAAGAACCTTANC
gcctccctcgcgccatcag NNNN NCAACGCGAAAAACCTTACC
gcctccctcgcgccatcag NNNN NCAACGCGCAGAACCTTACC
gcctccctcgcgccatcag NNNN NATACGCGARGAACCTTACC
gcctccctcgcgccatcag NNNN NCTAACCGANGAACCTYACC

3′ Primers

CGACAGCCATGCANCACCT
CGACAACCATGCANCACCT
CGACGGCCATGCANCACCT
CGACGACCATGCANCACCT

Fig. 3 Fused 5 Primer with 5 bp linker for distinguishing products of different PCR libraries and 3 amplification primers

remove sequences with mismatches to 5′ proximal amplification primers, sequences containing one or more N's within the V6-tag, sequences with fewer than 50 bases after the proximal primer, and sequences that extend more than 100 bp beyond the proximal primers. BLASTN (Altschul et al. 1990) (with nondefault parameters -q -1 and -G 1) and the EMBOSS program fuzznuc (Rice et al. 2000) (with nondefault parameter – mismatch = 3) identify complete, mismatched, and partial distal primers that arise from early sequence termination. Both the proximal and distal primers are trimmed from high-quality reads prior to database searches and similarity calculations. The 99.75% accuracy of V6 tag sequencing (Huse et al. 2007) is more than adequate to identify best matches to sequences in a reference database.

The tag sequences lack sufficient evolutionary information for the inference of phylogenetic trees, and the size of the data sets from a single run of the GSFLX exceeds the computational limitations of sequence alignment and tree-building algorithms. The reads are also too short for current implementations of Bayesian classifiers such as the Ribosomal Database Project (RDP) Classifier (Cole et al. 2005). As an alternative, a Global Assignment of Sequence Taxonomy (GAST) algorithm compares the tag sequences with a reference database of hypervariable homologs extracted from ~500,000 nearly full-length rRNA sequences. The tag sequence data sets are initially collapsed to unique sequences, and the number of times each sequence occurs is recorded. In the case of V6, each unique tag sequence serves as a BLAST query against a reference database of 59,830 nonidentical V6 sequences extracted from 195,344 nearly full-length bacterial rRNAs derived from multiple sources (Cole et al. 2005; Ludwig et al. 2004; Wheeler et al. 2006) and unannotated environmental RNA sequences, which are assigned taxonomic ranks using the Ribosomal Database Project II's Classifier (Cole et al. 2005). The classifier employs a "bootstrap" score, which is a measure of confidence in the taxon assignment. GAST nominally accepts taxonomic assignments for bootstrap values of 80%. If the bootstrap value is <80%, the taxonomic assignment moves to a higher classification level until an 80% or better bootstrap value is achieved. For example, if the genus assignment has a bootstrap value of 70%, but the family had a value of 85%, that sequence would be assigned only as far as family and not to genus. The RDP Classifier does not classify sequences below the genus level. Failure of the RDP classifier to make taxonomic assignments at any rank usually reflects the absence of an appropriate reference sequence in its training set.

Since BLAST uses a local alignment algorithm that does not ensure the identification of the most similar sequence in a reference data base, GAST collects the top BLAST hits (to a maximum of 150) and aligns the best matching sequences with the query sequence using the program MUSCLE (Edgar 2004) (with parameters -diags and -maxiters 2). GAST calculates the global distance from the sample tag to each of the aligned sequence tags by summing the number of insertions, deletions, and mismatches divided by the length of the tag. The nearly full-length sequence or sequences with the minimal global distance relative to the query is the top GAST match(es). The top BLAST hit generally agrees with the best GAST match(es); however, for 5–25% of tags the best GAST match is to a reference sequence with a lower BLAST score (Huse, manuscript in review). This process is more computationally intensive than BLAST score ratios (Rasko et al. 2005) but results in a better match than BLASTN alone in 5–10% of comparisons, depending on how many tag sequences are highly divergent from any reference sequence.

Each tag sequence query is assigned a taxonomy on the basis of its top GAST match(es). In the limited cases where a tag reference sequence has multiple taxonomies due to mapping to more than one nearly full-length rRNA sequence with different taxonomies, it is necessary to compile the taxonomy for all of these corresponding 16S sequences (with RDP bootstrap values +80). Starting at the genus level, if a consensus (defined as a two-thirds majority) of the full-length taxonomies has valid taxonomic assignments that agree to the genus level, that taxonomy is not assigned to the tag. If there is no consensus, the taxonomic assignment moves up one level to family. If there is no consensus at the family level, the assignment of taxonomy to the tag continues to proceed up the tree. Occasionally, a tag cannot be assigned a taxonomic descriptor at the domain level. This is most often due to RDP not assigning taxonomy with an adequate bootstrap value, rather than a tag mapping to full-length sequences from different domains. Tags that do not have any BLAST hits whatsoever are not assigned a taxonomic name. GAST uses a 66% majority, although other values or a distributional *vs.* strict percentage approach can be implemented.

5 Expanded Perspective of Microbial Diversity and the "Rare Biosphere"

The first massively parallel pyrosequencing studies of hypervariable sequences in rRNA genes targeted the V6 region. A total of ~120,000 tag sequences represented microbial population structures for six paired samples (sample pairs collected at the same coordinates but from different depths) from the meso- and bathypelagic realms at different locations in the North Atlantic Deep Water (NADW) loop of the ocean conveyor belt, and two samples of basalt-hosted diffuse hydrothermal vent fluids collected from the 1998 eruption zone of Axial Seamount, an active submarine volcano located on the Juan de Fuca Ridge in the northeast Pacific Ocean (Sogin et al. 2006). The rarefaction analyses (Fig. 4) of best GAST matches describe unprecedented levels of bacterial complexity for the eight marine samples, yet none has reached the curvilinear or plateau phase (Sogin et al. 2006). Furthermore, Table 1 shows there are as many 3,290 distinct OTUs (Site FS396) according to their best matches in GAST analyses but the tags report 8,699 unique sequences from the same sample. Many of these unique tags display imperfect matches to the same V6 sequences from nearly full-length rRNA sequences in the reference data base.

A viable alternative is to assess diversity independent of taxonomy. For example, the program DOTUR (Schloss and Handelsman 2005) identifies clusters of related sequences that equate to OTUs according to user-defined levels of similarity. These clusters represent evolutionary units, or phylotypes. The number of sequences that join a cluster defines that group's relative abundance. Microbial ecologists often employ a cluster of Small Subunit Ribosomal RNA (SSU) sequences that differ by no more than 3% as a first approximation of an ecologically or evolutionarily relevant phylotype,

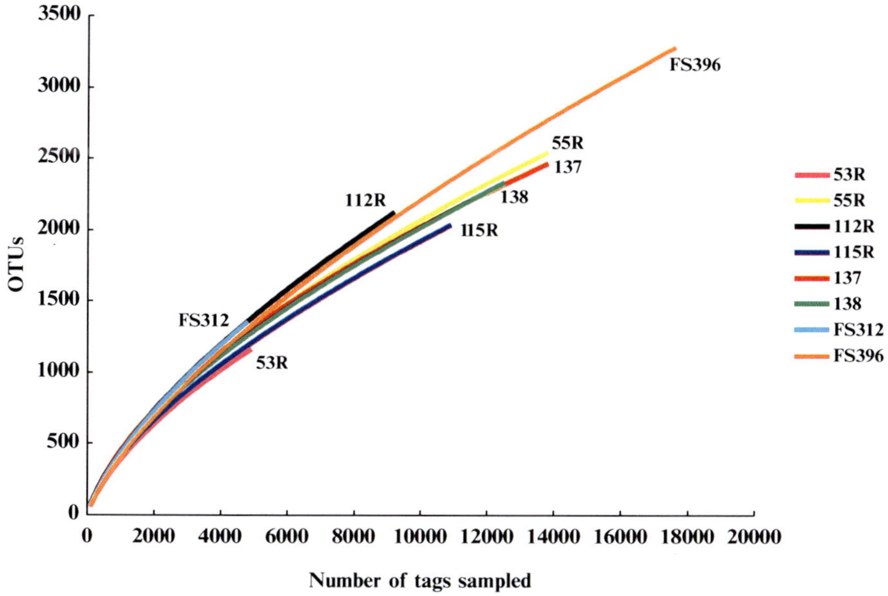

Fig. 4 Rarefaction analysis for each sample based upon best matches against the v6refdb. The frequency of observed best matches to the v6refdb (OTUs) for each site was used to calculate rarefaction curves with the program Analytic Rarefaction 1.3 (http://www.uga.edu/~strata/software/)

Table 1 Data summary and Phylotype OTUs

Sample ID	53R	55R	112R	115R	137	138	FS312	FS396
Total reads	6,505	18,439	12,916	14,731	18,137	18,451	6,605	22,994
Trimmed tags[a]	5,000	13,902	9,282	11,005	13,907	14,374	4,835	17,666
Unique tags[b]	2,656	7,187	5,752	5,777	6,752	7,168	2,769	8,699
OTUs[c]	1,184	2,555	2,135	2,049	2,480	2,550	1,362	3,290

[a]Number of reads remaining after removal of primers and low quality data
[b]Distinct sequences within set of trimmed tags
[c]Calculated by comparing each tag to reference V6 dataset, combining tags with same nearest reference combination

understanding that this is an oversimplification but that it is as likely to overestimate as to underestimate species diversity. The clustering of V6 regions according to the criteria of 3% sequence diversity performs very similarly to 3% clustering of the entire gene (unpublished). The original implementation of DOTUR does not accommodate tens of thousands of tag sequences. Similarly, no programs exist that can compute an alignment of hundreds of thousands of sequences, and the resulting distance matrix would be more than a terabyte in size. Using relational database tools, it is possible to compress tag sequence data to unique sequences prior to alignment and clustering, then to expand these clusters to contain all tags so that the cluster sizes can be used for calculating diversity and rarefaction. However, the limit of these analyses is still on the order of 37,000 unique tag sequences from a total data set of 725,000 sequences. Table 2 presents a DOTUR analysis of the NADW and Axial Sea

Mount tags. The diversity estimates for each of the samples exceeds all prior estimates of marine microbial diversity, with FS396 containing more than 25,000 OTUs in a liter of sea water.

More recently, this observation of microbial diversity was extended by attempting an exhaustive description of bacterial and archaeal population structures in the deep marine biosphere (Huber et al. 2007). Nearly one million tags were collected at the two low-temperature diffuse flow vents, Marker 52 (FS312) and Bag City (FS396), which are less than 3 km apart on Axial Seamount. These sites have significantly different chemical compositions and appearance. Figure 5 shows that we have almost exhaustively sampled the archaeal populations, which have a total diversity of just over 27,000 kinds of microbes. In contrast, bacterial population diversity is far from reaching a plateau. Nonparametric analysis indicates that there are more than 40,000 different kinds of bacteria. Microbial diversity is much greater than

Table 2 Similarity-based OTUs and species richness estimators[a]

Dist		0.03			0.05			0.10		
Sample	Reads	OTU	ACE	Chao1	OTU	ACE	Chao1	OTU	ACE	Chao1
53R	5,000	1,946	7,247	6,997	1,732	5,616	5,288	1,316	3,351	3,018
55R	13,902	5,266	19,235	18,191	4,673	14,959	14,209	3,644	9,618	9,080
112R	9,282	4,241	16,002	13,772	3,770	12,341	10,870	2,958	8,011	7,108
115R	11,005	4,000	12,767	11,296	3,413	9,189	8,585	2,457	5,553	5,448
137	13,907	4,554	13,698	11,991	3,866	9,734	8,705	2,708	5,353	5,181
138	14,374	5,136	18,656	16,600	4,508	13,852	12,424	3,293	7,596	6,941
FS312	4,835	1,941	5,599	5,482	1,681	4,233	4,080	1,227	2,466	2,346
FS396	17,666	6,326	23,315	20,949	5,573	18,003	16,889	4,291	11,520	10,567

[a]Determined by DOTUR

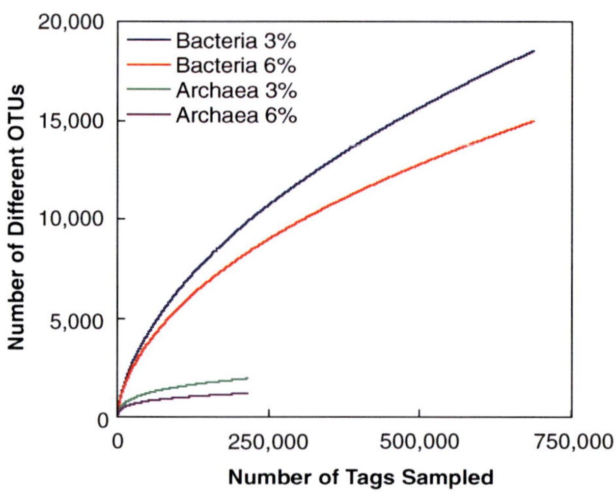

Fig. 5 Rarefaction curves for total bacterial and archaeal communities at the two sampling sites Marker 52 (FS312) and Bag city (FS396) at 3 and 6% difference levels

previous estimates from conventional molecular techniques. In traditional molecular studies, dominant populations have masked the detection of low-abundance organisms, their overwhelming diversity, and their individual distribution patterns in both terrestrial and aquatic environments. The relative abundance of different OTUs in the tag sequence data set varied by more than 3 orders of magnitude.

A relatively small number of tag sequences represented dominant bacterial populations in samples from the diffuse flow and deep waters of the North Atlantic. The vast majority (75%) of the tag sequences are very similar one another and to entries in the reference database. Underlying the major populations is a broad distribution of distinct bacterial taxa that represent extraordinary diversity. Approximately 25% of the "total tags" are identical to a sequence in our reference database, 40% are no more than 3% different and ~75% are no more than 10% different from any entry in the reference database. On the other hand, analyses of "unique reads" reveal that 80% are more than 10% different from a closest match in the reference database and, as indicated by the distribution of "all tags", the most divergent tag sequences represent very low abundance OTUs. These highly divergent low-abundance organisms constitute a "rare biosphere" that is largely unexplored. Some of its members might serve as keystone species within complex consortia, others might simply be the products of historical ecological change with the potential to become dominant in response to shifts in environmental conditions that favor their growth. Because we know so little about the global distribution of members of the rare biosphere, it is not yet possible to know whether they represent specific biogeographical distributions of bacterial taxa, functional selection by particular marine environments, or cosmopolitan distribution of all microbial taxa – the "everything is everywhere" hypothesis.

6 Massively Parallel DNA Sequencing and Future Surveys of Microbial Population Structures

In traditional molecular studies, dominant populations have masked the detection of the vast majority of low-abundance OTUs that make up the long tail of taxon-rank distribution curves, their overwhelming genetic diversity, and their individual distribution patterns. The sum of all different kinds of organisms in the rare biosphere represents a significant fraction of complete microbial communities. Massively parallel, high-throughput DNA sequencing provides a tool for detecting the presence of these organisms in different environmental contexts and monitoring shifts in relative abundance of minor populations. If microbial ecologists were to survey a sufficient number of distinct sites, it will be possible to identify populations that might be persistently rare, to detect low-abundance microbial taxa that reflect biogeographical distribution patterns from endemic sites, and to monitor the relative abundance of "seed" populations that respond to ecological change. These measurements afforded by new sequencing technology will be essential for developing a comprehensive understanding of how microbial communities shift and evolve over different spatial and temporal scales.

References

Altschul SF, Gish W, Miller W, Meyers EW, Lipman DJ (1990) Basic local alignment search tool. J Mol Biol 215:403–410

Atlas RM, Bartha R (1993) Microbial ecology. Fundamentals and applications. Benjamin/Cummings, Redwood City, CA

Butler J, MacCallum I, Kleber M, Shlyakhter IA, Belmonte MK, Lander ES, Nusbaum C, Jaffe DB (2008) ALLPATHS: de novo assembly of whole-genome shotgun microreads. Genome Res 18:810–820

Chao A (1984) Non-parametric estimation of the number of classes in a population. Scand J Stat 11:265–270

Chao A (1992) Estimating the number of classes via sample coverage. J Am Stat Assoc 87:210–217

Chao A, Ma MC, Yang K (1993) Stopping rules and estimation for recapture debugging with unequal failure rates. Biometrika 80:193–201

Cole JR, Chai B, Farris RJ, Wang Q, Kulam SA, McGarrell DM, Garrity GM, Tiedje JM (2005) The Ribosomal Database Project (RDP-II): sequences and tools for high-throughput rRNA analysis. Nucleic Acids Res 33:D294–D296

DeSantis TZ, Stone CE, Murray SR, Moberg JP, Andersen GL (2005) Rapid quantification and taxonomic classification of environmental DNA from both prokaryotic and eukaryotic origins using a microarray. FEMS Microbiol Lett 245:271–278

DeSantis TZ, Brodie EL, Moberg JP, Zubieta IX, Piceno YM, Andersen GL (2007) High-density universal 16S rRNA microarray analysis reveals broader diversity than typical clone library when sampling the environment. Microb Ecol 53:371–383

Edgar RC (2004) MUSCLE: multiple sequence alignment with high accuracy and high throughput. Nucleic Acids Res 32:1792–1797

Edwards RA, Rodriguez-Brito B, Wegley L, Haynes M, Breitbart M, Peterson DM, Saar MO, Alexander S, Alexander ECJ, Rohwer F (2006) Using pyrosequencing to shed light on deep mine microbial ecology under extreme hydrogeologic conditions. BMC Genomics 7:57 doi:10.1186/1471–2164–7–57

Felsenstein J (1989) PHYLIP – phylogeny inference package (version 3.2). Cladistics 5:164–166

Goldberg SMD, Johnson J, Busam D, Feldblyum T, Ferriera S, Friedman R, Halpern A, Khouri H, Kravitz SA, Lauro FM, Li K, Rogers Y-H, Strausberg R, Sutton G, Tallon L, Thomas T, Venter E, Frazier M, Venter JC (2006) A Sanger/pyrosequencing hybrid approach for the generation of high-quality draft assemblies of marine microbial genomes. PNAS 103:11240–11245

Guerrero R (2001) Bergey's manuals and the classification of prokaryotes. Int Microbiol 4:103–109

Gutell RR, Weiser B, Woese CR, Noller HF (1985) Comparative anatomy of 16-S-like ribosomal RNA. Prog Nucleic Acid Res Mol Biol 32:155–216

Higgins DG (1994) CLUSTAL V: multiple alignment of DNA and protein sequences. Methods Mol Biol 25:307–318

Huber JA, Mark Welch DB, Morrison HG, Huse SM, Neal PR, Butterfield DA, Sogin ML (2007) Microbial population structures in the deep marine biosphere. Science 318:97–100

Huse SM, Huber JA, Morrison HG, Sogin ML, Mark Welch DB (2007) Accuracy and quality of massively-parallel DNA pyrosequencing. Genome Biol 8:R143

Ley RE, Harris JK, Wilcox J, Spear JR, Miller SR, Bebout BM, Maresca JA, Bryant DA, Sogin ML, Pace NR (2006) Unexpected diversity and complexity of the Guerrero Negro hypersaline microbial mat. Appl Environ Microbiol 72:3685–3695

Ludwig W, Strunk O, Westram R, Richter L, Meier H, Yadhukumar, Buchner A, Lai T, Steppi S, Jobb G, Forster W, Brettske I, Gerber S, Ginhart AW, Gross O, Grumann S, Hermann S, Jost R, Konig A, Liss T, Lussmann R, May M, Nonhoff B, Reichel B, Strehlow R, Stamatakis A, Stuckmann N, Vilbig A, Lenke M, Ludwig T, Bode A, Schleifer KH (2004) ARB: a software environment for sequence data. Nucleic Acids Res 32:1363–1371

Margulies M, Egholm M, Altman WE, Attiya S, Bader JS, Bemben LA, Berka J, Braverman MS, Chen Y-J, Chen Z, Dewell SB, Du L, Fierro JM, Gomes XV, Godwin BC, He W, Helgesen S,

Ho CH, Irzyk GP, Jando SC, Alenquer MLI, Jarvie TP, Jirage KB, Kim J-B, Knight JR, Lanza JR, Leamon JH, Lefkowitz SM, Lei M, Li J, Lohman KL, Lu H, Makhijani VB, McDade KE, McKenna MP, Myers EW, Nickerson E, Nobile JR, Plant R, Puc BP, Ronan MT, Roth GT, Sarkis GJ, Simons JF, Simpson JW, Srinivasan M, Tartaro KR, Tomasz A, Vogt KA, Volkmer GA, Wang SH, Wang Y, Weiner MP, Yu P, Begley RF, Rothberg JM (2005) Genome sequencing in microfabricated high-density picolitre reactors. Nature 437:376–380

Medlin L, Elwood HJ, Stickel S, Sogin ML (1988) The characterization of enzymatically amplified eukaryotic 16S-like rRNA-coding regions. Gene 71:491–499

Moore M, Dhingra A, Soltis P, Shaw R, Farmerie W, Folta K, Soltis D (2006) Rapid and accurate pyrosequencing of angiosperm plastid genomes. BMC Plant Biol 6:17

Pace NR (1997) A molecular view of microbial diversity and the biosphere. Science 276:734–740

Pace NR, Stahl DA, Lane DJ, Olsen GJ (1985) Analyzing natural microbial populations by rRNA sequences. ASM News 51:4–12

Pedros-Alio C (2006) Marine microbial diversity: can it be determined? Trends Microbiol 14:257–263

Rasko DA, Myers GS, Ravel J (2005) Visualization of comparative genomic analyses by BLAST score ratio. BMC Bioinformatics 6:2

Rice P, Longden I, Bleasby A (2000) EMBOSS: the european molecular biology open software suite. Trends Genet 16:276–277

Ronaghi M, Uhlen M, Nyren P (1998) A sequencing method based on real-time pyrophosphate. Science 281:363–365

Schloss PD, Handelsman J (2005) Introducing DOTUR, a computer program for defining operational taxonomic units and estimating species richness. Appl Environ Microbiol 71:1501–1506

Smith LM, Sanders JZ, Kaiser RJ, Hughes P, Dodd C, Connell CR, Heiner C, Kent SB, Hood LE (1986) Fluorescence detection in automated DNA sequence anlaysis. Nature 321:674–679

Sogin ML, Morrison HG, Huber JA, Mark Welch D, Huse SM, Neal PR, Arrieta JM, Herndl GJ (2006) Microbial diversity in the deep sea and the underexplored "rare biosphere". PNAS 103:12115–12120

Sogin SJ, Sogin ML, Woese CR (1971) Phylogenetic measurement in procaryotes by primary structural characterization. J Mol Evol 1:173–184

Swofford DL (2001) PAUP*. Phylogenetic analysis using parsimony (*and other methods), 4.0b8 edn. Sinauer, Sunderland, MA

Van de Peer Y, Chapelle S, De Wachter R (1996) A quantitative map of nucleotide substitution rates in bacterial rRNA. Nucleic Acids Res 24:3381–3391

Wheeler DL, Barrett T, Benson DA, Bryant SH, Canese K, Chetvernin V, Church DM, DiCuccio M, Edgar R, Federhen S, Geer LY, Helmberg W, Kapustin Y, Kenton DL, Khovayko O, Lipman DJ, Madden TL, Maglott DR, Ostell J, Pruitt KD, Schuler GD, Schriml LM, Sequeira E, Sherry ST, Sirotkin K, Souvorov A, Starchenko G, Suzek TO, Tatusov R, Tatusova TA, Wagner L, Yaschenko E (2006) Database resources of the National Center for Biotechnology Information. Nucleic Acids Res 34:D173–D180

Whitman WB, Coleman DC, Wiebe WJ (1998) Prokaryotes: the unseen majority. Proc Natl Acad Sci USA 95:6578–6583

Woese CR, Fox GE (1977) Phylogenetic structure of the prokaryotic domain: the primary kingdoms. Proc Natl Acad Sci USA 74:5088–5090

Woese CR, Sogin ML, Sutton LA (1974) Procaryote phylogeny. I. Concerning the relatedness of Aerobacter aerogenes to Escherichia coli. J Mol Evol 3:293–299

Woese CR, Kandler O, Wheelis ML (1990) Towards a natural system of organisms: proposal for the domains Archaea, Bacteria and Eucarya. Proc Natl Acad Sci USA 87:4576–4579

Yooseph S, Sutton G, Rusch DB, Halpern AL, Williamson SJ, Remington K, Eisen JA, Heidelberg KB, Manning G, Li W, Jaroszewski L, Cieplak P, Miller CS, Li H, Mashiyama ST, Joachimiak MP, van Belle C, Chandonia JM, Soergel DA, Zhai Y, Natarajan K, Lee S, Raphael BJ, Bafna V, Friedman R, Brenner SE, Godzik A, Eisenberg D, Dixon JE, Taylor SS, Strausberg RL, Frazier M, Venter JC (2007) The Sorcerer II Global Ocean Sampling expedition: expanding the universe of protein families. PLoS Biol 5:e16

Detection and Characterization of Uncultivated Microorganisms Using Microarrays

Terry J. Gentry, Zhili He, and Jizhong Zhou (✉)

Contents

Abstract Microarrays have unprecedented potential for the high-throughput detection and characterization of uncultivated microorganisms. Several different types of arrays have been developed or adapted for the interrogation of microbial genomes and monitoring microbial population dynamics and/or activity in relation to various microbial processes such as bioremediation and biogeochemical cycling.

J. Zhou and Z. He
Institute for Environmental Genomics, Department of Botany and Microbiology, University of Oklahoma, 101 David L. Boren Blvd., Norman, OK 73019, USA
e-mail: jzhou@ou.edu

T.J. Gentry
Department of Soil & Crop Sciences, Texas A&M University, 2474 TAMU, 370 Olsen Blvd., College Station, TX 77843

Microbiol Monogr, doi:10.1007/7171_2008_6

Even though the number of such microarray studies has increased dramatically over the last few years, microarray analysis of uncultivated microorganisms still poses several challenges including a lack of sequence information for many organisms and related issues regarding the sensitivity and specificity of detection. As research continues to address these difficulties, and with further technological advances, microarrays will undoubtedly find even broader application to the investigation of uncultivated microorganisms, thus greatly increasing our understanding of the genetics, physiology, and distribution of these heretofore largely uncharacterized microorganisms.

1 Introduction

Research over the last few years has begun to reveal the incredible diversity of microorganisms that inhabit many environments, with one study suggesting as many as a million distinct genomes per gram of soil (Gans et al. 2005). While there is some debate over the accuracy of that specific estimation (Bunge et al. 2006; Volkov et al. 2006), it is now well established that there are enormous numbers and diversity of environmental microorganisms that we know little to nothing about – largely because the inability to cultivate these organisms has precluded their phenotypic characterization (Handelsman 2004). The development of molecular methods (e.g., polymerase chain reaction, PCR) has reduced the need for cultivation as a prerequisite for microbial characterization and has resulted in the discovery of many novel organisms; however, most of these methods are relatively of low throughput and reveal only limited information, usually phylogenetic, about the organisms they detect. Other methods such as metagenomic sequencing can reveal enormous amounts of information about uncultivated microorganisms or communities but are typically only used on a small number of samples. Microarray-based analyses have the potential to bridge this analytical gap.

Microarrays have unprecedented capability as a specific, sensitive, quantitative, and high-throughput technology for the detection and characterization of uncultivated microorganisms. Microarrays are currently available that can simultaneously detect thousands of different target organisms or genes, and this capacity continues to increase as the technology rapidly progresses (Brodie et al. 2006; He et al. 2007). Novel types of microarrays are continually being developed, but most arrays can be classified in one of five groups on the basis of the design and targets of the array: phylogenetic oligonucleotide arrays (POAs), functional gene arrays (FGAs), community genome arrays (CGAs), metagenomic arrays (MGA), and whole-genome open reading frame arrays (WGA). These microarrays are rapidly being developed and/or adapted to environmental microbiology-related research with nearly 40 such papers being published in 2006 alone (Wagner et al. 2007). Researchers have used different approaches for microarray-based analyses (Fig. 1) to investigate pertinent topics ranging from the elucidation of microorganisms involved in bioremediation

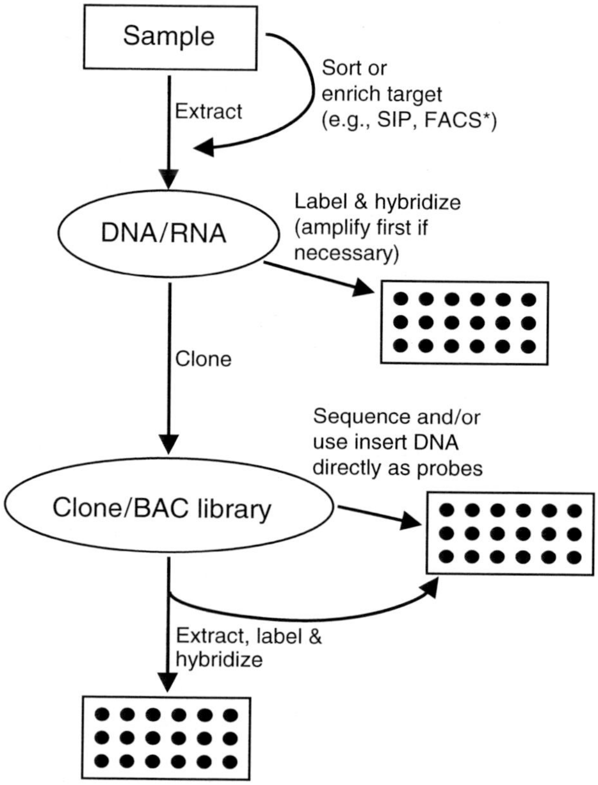

Fig. 1 Possible approaches for microarray analysis of uncultivated microorganisms (**SIP* Stable isotope probing, *FACS* Fluorescence-activated cell sorting)

and biogeochemical cycling processes to the detection of potential bioterrorism agents. The use of microarrays for studying uncultivated microorganisms still poses several challenges with respect to target sequence information, sensitivity, specificity, and data analysis, but applications will undoubtedly continue to increase as advances in microarray technology are made.

2 Types of Microarrays

2.1 *Phylogenetic Oligonucleotide Arrays*

Most POAs are based on specific regions of the ssu-rRNA (i.e., 16S rRNA) gene (Table 1). The rRNA gene has several characteristics that make it useful as a phylogenetic marker: e.g., (1) it exists in all organisms; (2) it appears to be rarely transferred between organisms; (3) it contains both conserved and variable regions of sequence;

Table 1 Microarrays for characterization of uncultivated microorganisms[a]

Property	*Type of Array*				
	POA	FGA	CGA	MGA	WGA
Probe template	Ribosomal rRNA genes	Functional genes	Whole genome	Environmental DNA or RNA	Open reading frames in whole genome
Probe target	rRNA or rDNA	mRNA or DNA	DNA	DNA or RNA	DNA or mRNA
Probe length	~18–25 nt	~50–70 nt oligos or ~200–1,000 nt PCR products	Whole genome	~1,000 nt	~25 nt oligos or ~200–3,000 nt PCR products
Information provided	Phylogenetic	Phylogenetic and functional	Phylogenetic	Functional	Phylogenetic and functional
Specificity	Species level or single nucleotide difference	< 80–90% sequence homology	Species – strain	≥Strain	Strain or single nucleotide difference
Sensitivity (ng of DNA)	~500	~1–8 (~0.01[b])	~0.2 (~0.1[b])	Undetermined	~0.1 (~0.04[b]) [PCR-based probes]

POA Phylogenetic oligonucleotide array, *FGA* Functional gene array, *CGA* Community genome array, *MGA* Metagenomic array, *WGA* Whole genome open reading frame array

Change Poa, fga, cga, mga, and wga to all caps like the table header and the Figure 1 caption.

[a]Adapted from (Zhou 2003)

[b]Quantitative detection limits using whole-community genome amplification methods are indicated in parentheses (Wu et al. 2006)

and (4) a vast amount of sequences exist in public databases such as the Ribosomal Database Project (RDP) (Cole et al. 2007).

One challenge for POAs targeting rRNA; (or DNA) is the formation of secondary structures in the target, which could result in decreased binding to the arrayed probes and thus reduce detection sensitivity. Several software programs exist that can identify self-complementarity in oligonucleotide probes, but it is difficult to predict the effects of target secondary structure on hybridization efficiency. Since the secondary structure of rRNA occurs in specific regions, it is possible to reduce the problem of target accessibility by designing probes complementary to regions less likely to form secondary structures. However, in contrast to the nonspecific binding of oligonucleotide probes to rRNA targets (false positives), which may be eliminated by stringent hybridization conditions, it is more difficult to eliminate instances where the probe and target cannot bind as a consequence of secondary structure in the target molecule (false negatives) (Peplies et al. 2003). One potential option to reduce this problem is to fragment or shear the target prior to hybridization (Chandler et al. 2003; Liu et al. 2007). Another potential option is the use of oligonucleotide helper probes designed to disrupt local secondary structure in the target by binding the target adjacent to the probe binding site (Peplies et al. 2003). This approach, however, can potentially be problematic if the helper probe binds too closely to either end of the capture probe binding site (Chandler et al. 2003). Furthermore, reduction of secondary structure in one region may result in the formation of secondary structure(s) in other regions of the target, potentially affecting the binding sites of other probes (Peplies et al. 2003). It is also difficult to use the helper probe approach with high-density arrays since the helper probes also need to have a level of specificity similar to the capture probes. Recent research suggests that it may be possible to overcome the limitations imposed on probe design by the secondary structure of the rRNA target as long as the thermodynamic affinity of the probe for the target is sufficient and the incubation period of the hybridization is extended to optimize the kinetics of target unfolding and probe binding (Yilmaz and Noguera 2004; Yilmaz et al. 2006).

The conserved nature of rRNA genes may necessitate the use of short oligonucleotides (~20-mers) as POA probes, especially for high density arrays, in order to achieve the required level of specificity. A common POA format for high-density arrays consists of using multiple probes that perfectly match a specific target along with corresponding probes containing a single mismatch, usually at the central position (Wilson et al. 2002; El Fantroussi et al. 2003; Peplies et al. 2003; Urakawa et al. 2003; Brodie et al. 2006; Brodie et al. 2007). Larger signal intensity for a perfectly matched probe relative to the mismatched probe indicates detection of the target sequence.

The POAs are among the most commonly used microarrays owing to the well-established use of rRNA as a tool for characterizing microbial communities. Researchers have used POAs to examined microbial populations in a diverse set of environments including air (Brodie et al. 2007), lake water and sediments (Rudi et al. 2000; Castiglioni et al. 2004), estuary sediments (El Fantroussi et al. 2003) and enrichments (Koizumi et al. 2002), soil (Brodie et al. 2006), activated sludge (Adamczyk et al. 2003), and hypersaline cyanobacterial mats (Loy et al. 2002). The largest POA developed to date, dubbed the PhyloChip, used the perfect match/mismatch approach and contained over half million 25-mer oligonucleotide probes (Brodie et al. 2006). Approximately 300,000 of the probes were perfect match/mismatch probes targeting 16S rRNA genes and the rest were for image orientation, controls, or other purposes. The PhyloChip was designed on the basis of >30,000 16S rRNA gene sequences. These sequences were grouped into ~9,000 operational taxonomic units (OTUs). The researchers attempted to design a set of 11 or more probes specific to each OTU, which were then screened for potential cross-hybridization. Detection was considered positive if (1) the perfect match probe signal was >1.3 times the mismatch probe signal and (2) the difference in the signal between the two probes (perfect match – mismatch) was at least 500 times greater than the squared noise value. The PhyloChip has been successfully used to characterize microbial communities in urban aerosols and to monitor subsurface microbial populations during in situ reduction of uranium (discussed in more detail in a later section) (Brodie et al. 2006, 2007).

2.2 *Functional Gene Arrays (GeoChips)*

In contrast to POAs, which are primarily designed on the basis of phylogenetic genes (e.g., rRNA), FGAs are based on genes that encode for specific functional processes or phenotypic characteristics. (While rRNA genes are technically functional genes critical to microbial growth, the term *functional gene* in this context has traditionally been used to indicate genes that encode proteins that directly impact some externally measurable process besides cell growth.) This enables FGAs to both provide information on the genetic capacity for and/or activity of a specific microbial process and also provide a level of phylogenetic classification. However, FGAs are not as well suited as POAs for the design of broad-group and universal probes because of the variability of many functional genes. Therefore, individual FGA probes are typically designed to detect only specific or small groups of organisms.

Several different groups of genes encoding key enzymes in various metabolic processes have been used for FGAs, including those involved in biogeochemical cycles (Wu et al. 2001; Cho and Tiedje 2002; Bodrossy et al. 2003; Taroncher-Oldenburg et al. 2003; Jenkins et al. 2004; Rhee et al. 2004; Steward et al. 2004; Stralis-Pavese et al. 2004; Tiquia et al. 2004; He et al. 2007), contaminant remediation (Cho and Tiedje 2002; Denef et al. 2003; Dennis et al. 2003; Rhee et al. 2004; He et al. 2007), and pathogen detection/characterization (Call et al. 2003; Grimm et al. 2004; Korczak et al. 2005; Perreten et al. 2005). Important characteristics that any gene should possess in order to be potentially be included in an FGA include that (1) it encodes an important enzyme or protein in the process of interest; (2) it is evolutionarily conserved yet has enough sequence divergence in different organisms to allow probe design for individual species; and (3) it has substantial sequence data from various isolates and environmental samples available in public databases. Because of enormous environmental sequence variability, it may be beneficial in many instances to initially do clone libraries for the gene and environment of interest in order to verify that the sequences used for FGA design adequately represent sequence diversity in a specific set of samples. Caution should be exercised when attempting to link FGA results based on "key" genes with the capacity for a specific function since the presence of a particular gene may not indicate the presence of all the genes necessary to complete the process of interest.

The FGA probes can be either PCR amplicons of various functional genes (Wu et al. 2001; Cho and Tiedje 2002) or synthesized oligonucleotides (Denef et al. 2003; Rhee et al. 2004). One advantage of PCR amplicon probes is that they can be constructed using conserved primers designed from known sequence data in other organisms and therefore do not require prior sequence information for the specific target organism/sample. It can, however, be practically impossible to obtain all the requisite isolates and environmental clones from their divergent sources in order to produce a comprehensive PCR amplicon FGA. In contrast, synthesized, oligonucleotide probes can be designed directly from available sequence data thereby eliminating the need to physically obtain (and maintain) the original isolates or clones. Oligonucleotide probes are also usually much shorter, being designed on the basis of only a portion of a particular gene. This potentially enables the design of better probes by allowing the avoidance of highly conserved regions or problematic regions of a gene. Additional probe design criteria also impact the specificity and sensitivity of both PCR amplicon and oligonucleotide probes as discussed in subsequent sections.

The development of comprehensive FGAs is currently limited only by the availability of essential isolates and/or sequence data and the capital resources needed for array construction. The largest FGA constructed to date, dubbed the GeoChip, contains over twenty-four thousand 50-mer oligonucleotide probes representing >10,000 genes involved in the C, N, and S cycles; organic contaminant degradation; and metals resistance and reduction (He et al. 2007). The GeoChip has been already been used in a variety of studies that delineated the microbial populations involved in bioremediation of uranium (He et al. 2007) and polychlorinated biphenyls (Leigh et al. 2007), examined the diversity of C and N cycling genes across an Antarctic transect (Yergeau et al. 2007), and interrogated the genomes of nickel-resistant bacteria (Van Nostrand et al. 2007).

2.3 *Community Genome Arrays*

The community genome array(CGA) is similar to membrane-based reverse sample genome probing (Greene and Voordouw 2003), but instead uses a nonporous hybridization surface and fluorescence-based detection. These features enable high throughput analyses but reduce the sensitivity of detection. Wu et al. (2004) developed a CGA which contained the genomic DNA of 67 different bacteria including α-, β-, and γ-*Proteobacteria* and Gram-positive bacteria. The array contained one species' genome per spot and could achieve species- to-strain-level differentiation depending on the hybridization conditions. The primary disadvantage of the CGAs designed so far is that only cultivated microorganisms have been included. However, with recent advances in the creation of large-insert, metagenomic libraries, it may also be possible to use DNA from uncultivated organisms for microarray construction.

2.4 *Metagenomic Arrays*

As evidenced by the publication of this book, the vast majority of environmental microorganisms are yet to be, and possibly cannot be, isolated using current techniques. This has led to recent development of the field of metagenomics (see Miao and Davies (2008)) (Handelsman 2004). Metagenomic approaches have been used to sequence entire communities in an acid mine drainage site (Tyson et al. 2004) and a portion of a more complex community from the Sargasso Sea (Venter et al. 2004). While it is currently difficult to assemble even the dominant genomes from microbially diverse sites such as surface soils and sediments (Torsvik and Ovreas 2002; Venter et al. 2004; Tringe et al. 2005), the combination of microarray and metagenomic technologies has the potential to reveal considerable information on these yet-uncultivated microorganisms.

MGAs are still in the early stages of development but have enormous potential for characterizing both the genetic capacity and activity of microbial communities. One of the few MGAs published so far was based on ~1-kb inserts amplified from 672 cosmids derived from a groundwater enrichment and was used as a high-throughput method to screen groundwater isolates and community DNA (Sebat et al. 2003). While the above array contained probes that were only ~1 kb, larger fragments from fosmid or bacterial artificial chromosome libraries could potentially be used to provide higher genomic throughput (Berry et al. 2003; Wu et al. 2004). Furthermore, it may be possible to create MGAs for measuring microbial activity if enough mRNA can be obtained to generate a cDNA library from which probes could be designed. The use of MGAs has great promise for the study of uncultivated microorganisms since one of the major current limitations for microarray analysis is the lack of sequence information.

2.5 Whole-Genome Open Reading Frame Arrays

WGAs contain probes for all (or at least many) of the open reading frames (ORFs) in a genome. Like FGAs, MGAs can be either based on PCR amplicon or oligonucleotide probes. One application of WGAs is to study the environmental transcriptomics of relatively simple systems – containing a single or few different types of organisms (Barnett et al. 2004). These arrays can also be used for comparative genomics with specific potential for studying lateral gene transfer and microevolution (Behr et al. 1999; Salama et al. 2000; Dong et al. 2001; Murray et al. 2001; Dziejman et al. 2002; Ochman et al. 2005; Ochman and Santos 2005). Custom WGAs can be designed, based only on sequence data, through companies such as NimbleGen Systems Inc. (http://www.nimblegen.com/products/cgr/index.html) for rapidly surveying microbial genomes through comparative genome sequencing via microarrays.

2.6 Isotope Arrays

While microarrays have unprecedented power for the high-throughput detection and characterization of microorganisms, their greatest utility lies perhaps in combined use with other complementary methods (Wagner et al. 2007). This approach can be used not only to validate microarray results but also to produce robust, synergistic tools for investigating microbial processes. One of the potentially most powerful combined approaches is the use of microarrays in conjunction with isotope techniques. By using labeled substrates (e.g., ^{13}C or ^{14}C), this system can achieve high-throughput differentiation between microorganisms which are actively involved in a process of interest and those that are not, based on incorporation of the labeled substrate by the active organisms. Both radioisotopes (e.g., ^{14}C) and stable isotopes (e.g., ^{13}C) have been successfully used for these purposes (Radajewski et al. 2000; Adamczyk et al. 2003; Polz et al. 2003; Wagner et al. 2006; Cebron et al. 2007; Leigh et al. 2007). One distinction between radioisotopes and stable isotopes is that methods are currently available to detect radioisotopes directly on an array. The use of the stable isotopes requires separation of labeled and unlabeled nucleic acids, usually by centrifugation, prior to microarray hybridization, which can limit the sensitivity of stable isotope detection. However, this limitation may be eliminated in the future if methods are developed for directly detecting ^{13}C on arrays.

2.7 *Other Types of Arrays*

Numerous other types of microarrays have been used for microbial characterization – primarily of cultivated isolates. Random or digested genomic fragments have been used as probes on arrays for applications where the genome sequences of the target organisms were unknown (Cho and Tiedje 2001; Parro and Moreno-Paz 2003; Kim et al. 2004). For some applications, mixed-genome arrays containing random gene fragments from multiple organisms may enable greater discrimination (Wan et al. 2007). Likewise, randomly selected oligonucleotide arrays (consisting of 47 nonamer probes randomly generated based on the *E. coli* K-12 genome) have been used to fingerprint bacteria (Kingsley et al. 2002). In another approach researchers generated an array based on a library of 9,600 rRNA clones (Bent et al. 2006). The rRNA inserts were PCR amplified, arrayed, and then hybridized with a series of different oligonucleotide probes to generate a fingerprint for each clone. These fingerprints were then used to classify new, unidentified clones.

As the technology continues to progress, new array formats will undoubtedly expand the utility of microarrays for analysis of uncultivated microorganisms. While advances in extraction and amplification methods are making microarray analysis of environmental RNA more practical, this is only an indirect measure of microbial activity. The development and adaption of protein arrays would potentially allow direct measurement of enzymatic expression in environmental samples (Duburcq et al. 2004; Ro et al. 2005; Bjornstad et al. 2006; Gregson et al. 2006). Additional research may ultimately improve microarray automation and make it possible to construct microarrays that are field-deployable and can achieve real-time characterization of microbial communities (Chandler and Jarrell 2004; Liu et al. 2004; Liu and Zhu 2005).

3 Applications of Microarrays

3.1 *Microbial Detection*

The high-throughput potential of microarrays can be advantageous for several applications where it is necessary to simultaneously detect many different microorganisms, such as assessing microbial diversity and/or biogeography or screening for potential bioterrorism agents (Table 2). In one of the most descriptive ecological applications of POAs to date, a POA containing 132 probes (18-mer) was used to characterize sulfate-reducing bacteria at four depths (ranging from 0 to 30 cm) in two acidic, low-sulfate fens (wetland soils) in Germany (Loy et al. 2004). The POA consisted of probes specific to the rRNAs of individual and groups of organisms, covering all known lineages of sulfate-reducing bacteria. The fens differed in iron content, vegetation, acidity, and to some degree, seasonal water saturation. The POA results indicated that stable sulfate-reducing populations varied little with depth within each of the two sites but were different between the sites. Members of the *Syntrophobacteraceae* were detected in the upper 30 cm of both sites, but *Desulfomonile* spp. were only found in one soil which also contained a more diverse sulfate-reducing community. These results were confirmed by direct PCR amplification with the appropriate group-specific rRNA primers and by the detection of the corresponding *dsrAB* genes from the samples. The same research group also developed a POA targeting all of the cultured and uncultured members of the *Rhodocyclales* (Loy et al. 2005). The array detected *Rhodocyclales* populations representing <1% of the total community, following *Rhodocyclales*-selective PCR amplification. The POA indicated the presence of several uncultured *Zoogloea*-, *Ferribacterium/Dechloromonas*-, and *Sterolibacterium*-like organisms in the activated sludge from an industrial wastewater treatment plant, which was corroborated by the results from a 16S rRNA gene clone library. The results also demonstrated that the *Rhodocyclales* community in the reactor, thought to represent the major denitrifiers in the system, had dramatically changed, possibly as a result of alterations in treatment plant operations.

One of the most comprehensive POAs developed so far contained 31,179 perfectly matched hierarchical 20-mer probes (with a corresponding number of single mismatch probes as negative, mismatch controls) targeting 1,945 prokaryotic and 431 eukaryotic sequences (Wilson et al. 2002). The array was used to investigate microorganisms collected from a 1.4 million liter air sample. The POA results generally agreed with those from an rRNA gene clone library, but could only resolve differences to the third level of phylogenetic rank, as defined by RDP, and could not identify individual species. Eight of 10 phylogenetic clusters detected by the array were represented in the rRNA gene clone libraries, and the organisms not detected had relatively low signals on the array. Approximately 7% of the clones were not detected by the POA, but these were from novel organisms not represented in the RDP or on the array. In contrast, there was no good correlation between the relative numbers of clones in each group and the signal intensity of that group detected by the array, indicating a potential limitation with respect to microbial quantitation with this system. Subsequent research has indicated that this array format does enable quantitative detection and may actually reveal more diversity than a normal-sized clone library (Desantis et al. 2005, 2007). A larger, more recent version of the array was used to monitor urban aerosols in two U.S. cities over a 17-week period (Brodie et al. 2007). The POA analysis indicated that the aerosols contained at least 1,800 diverse bacterial types including several organisms related to known pathogens and potential bioterrorism agents.

Table 2 Examples of studies that used microarrays for microbial characterization

Application	Sample	Type of array	Target	Amplification or selection	Results	Reference
Bioremediation	Soil	POA	rDNA	PCR	Demonstrated that populations of uranium-reducing bacteria were maintained following uranium reduction and were not the likely cause of subsequent uranium re-oxidation	Brodie et al. (2006)
Bioremediation	Groundwater	FGA	DNA	Whole-genome community amplification	Identified microorganisms whose population dynamics were correlated with in situ reduction of uranium in a contaminated aquifer	He et al. (2007)
Bioremediation	Mixed culture, reactor	FGA	mRNA	None	Detected increased expression of specific functional genes during degradation of a herbicide (2,4-D)	Dennis et al. (2003)
Bioremediation	Enrichment, soil	FGA	DNA, mRNA	None	Detection of naphthalene degradation genes in soils and those up-regulated in enrichments during degradation	Rhee et al. (2004)
Bioremediation	Soil	FGA	mRNA	Stable isotope probing	Identification of functional genes expressed during degradation of ^{13}C-labeled polychlorinated biphenyls	Leigh et al. (2007)
Biogeochemical cycling	Activated sludge	POA	rRNA	None	Combined with ^{14}C-based isotope techniques for parallel detection of ammonia-oxidizing bacteria and CO_2 fixation in nitrifying activated sludge	Adamczyk et al. (2003)
Biogeochemical cycling	Soil	POA	rDNA	PCR	Detection of sulfate-reducing bacteria in low-sulfate, acid fens	Loy et al. (2004)
Biogeochemical cycling	Soil	FGA	DNA	Stable isotope probing	Identified microorganisms actively consuming $^{13}CH_4$	Cebron et al. (2007)
Biogeochemical cycling	Soil	FGA	DNA	PCR	Linked differences in and methane oxidation with differences in vegetation and populations of specific groups of methanotrophs	Stralis-Pavese et al. (2004)
Environmental transcriptomics	River water	Other	mRNA	T7 RNA polymerase – amplification	Measured in situ gene expression for a dominant member of a highly acidic and metal contaminated ecosystem	Parro et al. (2007)
Metagenomic profiling	Riverine biofilm enrichment	MGA	DNA	None	Detection, isolation, and characterization of genes from uncultivated microorganisms	Sebat et al. (2003)
Microbial detection	Air	POA	rDNA	PCR	Characterization of airborne bacteria and detection of potential bioterrorism agents	Brodie et al. (2007)

POA Phylogenetic oligonucleotide array, *FGA* Functional gene array, *MGA* Metagenomic array

3.2 Microbial Population Dynamics and Activity

Several studies have used FGAs to investigate microbial involvement in biogeochemical processes including nitrogen fixation, nitrification, denitrification, sulfate reduction (Wu et al. 2001; Taroncher-Oldenburg et al. 2003; Jenkins et al. 2004; Steward et al. 2004; Tiquia et al. 2004), methane oxidation (Bodrossy et al. 2003; Stralis-Pavese et al. 2004), and remediation of organic and metal contaminants (Denef et al. 2003; Rhee et al. 2004; He et al. 2007; Leigh et al. 2007). In one of the largest-scale FGA applications to date, Stralis-Pavese et al. (2004) used an array containing 68 different 17- to 27-mer probes, primarily targeting the particulate methane monooxygenase (*pmoA*) genes of several methanotrophs, to investigate the impact of five different plant covers on methanotrophic activity in lysimeters under landfill-simulating conditions. The authors linked the methanotrophic community structure in the vegetated lysimeters, in which type II methanotrophs had a competitive advantage over type Ia methanotrophs, with increased methane oxidation relative to the nonvegetated lysimeters. Furthermore, the relative abundances of methanotrophs were lower in the lysimeters that did not receive biogas.

One of the first applications of the GeoChip FGA was to monitor groundwater microbial populations during in situ uranium reduction (He et al. 2007). The array detected several different c-type cytochrome and *dsrAB* (dissimilatory sulfite reductase) genes that were significantly correlated with uranium reduction. This included genes from *Geobacter*-type iron-reducing bacteria and *Desulfovibrio*-type sulfate-reducing bacteria, suggesting that these organisms actively or indirectly played a role in uranium reduction. Demonstrating the potential for biogeographical applications, Yergeau et al. (2007) used the GeoChip to examine soil microbial communities across an Antarctic latitudinal transect. The GeoChip revealed that cellulose degradation and denitrification genes were correlated with soil temperature, and N-fixation genes were linked to plots mainly vegetated by lichens. The GeoChip has also been used to detect biphenyl-degrading organisms in polychlorinated biphenyl–contaminated soil (Leigh et al. 2007).

One of the major potential benefits of FGAs is that they can also potentially be used to determine environmental gene expression by measuring mRNA. However, only a handful of studies have used FGAs for mRNA analysis to date (Dennis et al. 2003; Rhee et al. 2004; Bodrossy et al. 2006; Gao et al. 2007). Dennis et al. (2003) constructed a PCR amplicon FGA (271–1,300 bp fragments) containing probes for 64 genes including several from the 2,4-dichlorophenoxyacetic acid (2,4-D)-degradation pathway of *Ralstonia eutropha* JMP134 and related organisms. The FGA was used to test mixed cultures consisting of four isolates from a batch reactor treating pulp mill effluent and varying concentrations of *R. eutropha* JMP134. The cultures were amended with 2 mM of 2,4-D and incubated 6 h prior to mRNA extraction. Significant induction of 2,4-D degradation genes was detected from populations as low as 0.0037% (3.7×10^3 cells in 10^8 total community) to 3.7%, depending on the specific genes detected and sequence similarity of the probes that were used.

Rhee et al. (2004) used a 50-mer oligonucleotide FGA (a predecessor to the GeoChip that contained 1,662 probes) to determine both the presence and expression of naphthalene-degradation genes in soil enrichments. Previously contaminated soil was enriched with naphthalene. Four different naphthalene-degradation genes, three of which were from *Rhodococcus* spp., were detected at higher levels in the naphthalene-amended enrichment, based on DNA analysis. Likewise, FGA analysis of mRNA detected three different *Rhodococcus* sp. genes involved in naphthalene degradation that were upregulated (40- to 100-fold) in the naphthalene-amended enrichment, including two of the genes detected by DNA analysis. The results also indicated that other potential naphthalene-degrading organisms, whose genes were detected in the enrichments by the DNA hybridizations, were not the primary naphthalene degrading organisms under the tested conditions. An FGA has also been used to detect active methanotrophs in lysimeters simulating landfill covers (Bodrossy et al. 2006).

Most of the above studies that analyzed mRNA were relatively simple systems – mixed cultures or enrichments. In many cases, it remains challenging to use FGAs to analyze mRNAs from more complex environmental samples because of difficulties in extracting sufficient quantities of high-quality mRNA (Saleh-Lakha et al. 2005). While advances in extraction techniques (Hurt et al. 2001; Sessitsch et al. 2002; Burgmann et al. 2003) and newly available commercial kits (e.g., TruRNA from Atom Sciences, Inc. and FastRNA from Qbiogene) are helping with this process, it may be necessary to further purify many samples which may decrease mRNA yields. This can be especially problematic for low-biomass environments that already do not contain sufficient mRNA for FGA analysis. New approaches such as (1) the addition of a poly(A) tail to bacterial RNA for subsequent amplification (Botero et al. 2005) or (2) the amplification of whole community mRNAs using random primers with an attached T7 promoter sequence for subsequent amplification (Bodrossy et al. 2006; Moreno-Paz and Parro 2006; Gao et al. 2007) will likely expand the applications of microarray analyses for determinations of microbial activity in environmental samples.

Besides FGAs, other types of arrays also can be used to study different microbial processes. The PhyloChip POA containing >500,000 probes based on 16S rRNA gene sequences was used to examine microbial populations in subsurface material during uranium remediation (Brodie et al. 2006). PhyloChip analysis revealed that populations of metal-reducing bacteria (e.g., *Geobacteraceae*) were abundant during uranium reduction (induced by lactate biostimulation) and did not decrease during subsequent uranium re-oxidation. These results indicated that decreases in the populations of uranium-reducing bacteria were likely not responsible for the re-oxidation.

3.3 *Microbial Characterization*

Several different types of arrays (e.g., CGAs, FGAs, and WGAs) have been used to characterize microbial genomes. While many of the examples below are for isolated microorganisms, similar analyses could be done for large insert metagenomic DNA. The main use of CGAs is to determine the genomic relatedness of isolated bacteria (or metagenomic DNA) to each other and also the organisms represented on the array. A CGA that contained the entire genomic DNA of 67 different bacteria (one species' genome per spot) including α-, β-, and γ-*Proteobacteria* and Gram-positive bacteria – most being *Azoarcus*, *Pseudomonas*, or *Shewanella* spp. – was designed as a tool for detecting specific microorganisms within a natural microbial community (Wu et al. 2004). The CGA could achieve species- to strain-level differentiation depending on the hybridization temperature used. The CGA was also capable of correlating differences in microbial communities in different environmental samples with differences in biogeochemical and physical properties between the sites.

Similarly, WGAs can be used to characterize microbial isolates or metagenomic DNA. Dong et al. (2001) used a WGA containing 96% of the annotated ORFs in *Escherichia coli* K-12 to comparatively interrogate the genome of a related bacterium, *Klebsiella pneumoniae* 342. Only 70% of *E. coli* K-12 ORFs were found in *K. pneumoniae* 342. Both bacteria shared many highly conserved genes including those for energy, amino acid, and fatty acid metabolism along with cofactor synthesis, cell division, DNA replication, transcription, translation, transport, and regulatory proteins. In contrast, many *E. coli* K-12 hypothetical and putative regulatory proteins, chaperones, and enzymes, in addition to genes thought to have been acquired from phage, plasmids, or transposons via lateral transfer, were not found in *K. pneumoniae*. Other researchers have used a WGA to find evidence for horizontal gene transfer in *Shewanella* spp. (Murray et al. 2001).

Van Nostrand et al. (2007) used the GeoChip FGA to characterize the genomes of four nickel-tolerant bacteria isolated from contaminated sediment at the U.S. Department of Energy Savannah River Site in Aiken, SC. Phenotypic assays had indicated that the bacteria were resistant to several metals and antibiotics. Subsequent GeoChip analysis detected between 20 and 86 metal-related genes, including some for arsenic, chromium, copper, mercury, and nickel in the genomes of the different isolates.

4 Challenges and Limitations of Microarray Analysis

4.1 Reliance on Cultivated Organisms and/or Known Sequences

Many types of microarrays require isolated organisms or prior knowledge of target sequences for the construction of probes. This represents one of the greatest challenges for the analysis of environmental samples given the vast number of unknown DNA sequences in many samples. Microorganisms critical to the process of interest may be completely overlooked if they do not have corresponding probes on the array. Furthermore, probes that are designed to be specific to known sequences can also cross-hybridize to similar, unknown sequences from related or unrelated genes.

This can potentially result in either an underestimated signal due to weak binding of a slightly divergent sequence or a completely misleading signal due to binding of a nontarget sequence. In any case, microarray data from environmental samples should be cautiously interpreted.

4.2 Sensitivity

Another challenge for microarray analysis is the limited sensitivity of detection, with usually only dominant organisms being identified. Most microarrays are printed on nonporous, planar glass slide platforms because this enables high-density printing. However, this reduces detection sensitivity several orders of magnitude as compared to membrane-based hybridizations, most likely due to less probe being attached to the nonporous surfaces (Cho and Tiedje 2002). New slide chemistries, such as ultra thin or three-dimensional platforms, may ultimately help to increase detection sensitivity while maintaining the capacity for high-throughput printing (Guschin et al. 1997; Urakawa et al. 2003; Zhou et al. 2004; Hesse et al. 2006; Sunkara et al. 2007).

The choice of probe format can also impact sensitivity. Oligonucleotide probes have several advantages for probe design, but they are typically ~10- to 100-fold less sensitive than longer PCR amplicon or CGA probes (Wu et al. 2001; Denef et al. 2003; Rhee et al. 2004; Wu et al. 2004). For example, detection limits of 0.2 ng of target genomic DNA for a CGA (Wu et al. 2004), 1 ng for a PCR amplicon FGA (Wu et al. 2001), and 5–8 ng for a 50-mer oligonucleotide FGA (Rhee et al. 2004) have been reported for hybridization in the absence of background DNA. In the presence of background DNA, sensitivity is further reduced about 10-fold (Rhee et al. 2004; Tiquia et al. 2004; Wu et al. 2004). For one 50-mer FGA, this detection limit corresponded to ~10^7 cells or 5% of the total community, which is similar to other published studies (Cho and Tiedje 2002). One study recently directly compared the detection sensitivity of 50-, 60-, and 70-mer oligonucleotide probes with PCR amplicon probes (He et al. 2005a). The oligonucleotide probes had detection limits of 25–100 ng of genomic DNA, while the PCR amplicon probes had a detection limit of 5 ng of genomic DNA. These limits roughly corresponded to 10^7 and 10^6 gene copies for the oligonucleotide and PCR amplicon probes, respectively. The 70-mer probes produced results most similar to the PCR amplicons. The detection sensitivities of other arrays (MGAs and WGAs) will depend on the specific probe design.

Advances in nucleic acid labeling methods may also help to increase sensitivity (Denef et al. 2003; Steward et al. 2004; Zhou and Zhou 2004). One study reported that tyramide signal amplification labeling increased the signal intensity of a 70-mer FGA ~10-fold compared to commonly used Cy dye-labeling techniques (Denef et al. 2003). Use of new labels such as nanoparticles (Zhou and Zhou 2004) or even the development of label-free arrays may ultimately further enhance detection sensitivity (Thompson et al. 2005; Liu and Bazan 2006).

While the approaches discussed above may help to detect dominant members of relatively high biomass communities, improved methods are needed for detecting less abundant microorganisms. Potential approaches for selectively enriching specific targets include (1) PCR amplification of specific genes using conserved primers (Bodrossy et al. 2003), (2) use of capture techniques such as magnetic beads (Tsai et al. 2003), and (3) directed cell separation using fluorescence in situ hybridization and flow cytometry (Podar et al. 2007). For low-biomass samples, or small numbers of sorted cells, nonspecific amplification of whole community DNA can be used to generate sufficient material for microarray analysis (see Podar et al. 2008).

A whole-community genome amplification (WCGA) procedure has been developed, based on phi 29 DNA polymerase, that can amplify nanogram quantities of DNA to microgram quantities with a linear relationship between the starting template and the final product (Wu et al. 2006). The method was capable of quantitatively detecting 0.04–125 ng (r^2 = 0.65–0.99) of pure culture DNA using a WGA, 0.1–1,000 ng (r^2 = 0.91) of genomic DNA using a CGA, and 0.01–250 ng (r^2 = 0.96–0.98) of environmental DNA using a 50-mer FGA. The approach allowed representative detection of individual genes from 1 to 500 ng of initial environmental (groundwater) DNA template. Smaller amounts of template, as low as 10 fg, could be detected via WCGA, but this affected the representativeness of amplification.

4.3 *Specificity*

The specificity of microarray probes is governed by several factors including probe design and hybridization conditions. While probes can be designed to meet specific binding characteristics, at least theoretically, there is always potential for cross-hybridization if used with environmental samples that contain unsequenced microbial populations (Chandler and Jarrell 2004). This issue can be reduced by obtaining more sequence data (i.e., clone libraries) for specific samples prior to microarray analysis, but will likely remain a concern for most environments.

It is, however, possible to improve probe specificity for specific targets. This is especially true for oligonucleotide probes that can be designed to avoid conserved regions of genes or those producing stable secondary structures. Arrays can differentiate a single mismatch in a probe–target hybridization using shorter oligonucleotide probes (~20-mers) (Wilson et al. 2002; Urakawa et al. 2003; Zhou et al. 2004). One of the more common formats for POAs uses at least one probe for a given target that perfectly matches a target sequence and a corresponding probe(s) that contains one mismatched nucleotide, usually at a central position (Brodie et al. 2006). Detection of the target sequence is usually determined on the basis of greater signal intensity (using some predetermined level) for the perfectly matched probe relative to the mismatched probe. Owing to the potential for unexpected probe behavior (higher signal for the mismatched probe), several different sets of probes are commonly included on these arrays for each target sequence. Thermal dissociation curves for individual probe–target duplexes have been used to further improve the discriminating power of these arrays (Liu et al. 2001; El Fantroussi et al. 2003; Urakawa et al. 2003; Li et al. 2004; Eyers et al. 2006).

While short oligonucleotide probes are often preferred for POAs in order to achieve the desired level of specificity, it is possible to use longer oligonucleotide probes (~40- to 70-mers) for FGAs owing to the more variable nature of most functional genes. This allows for an increase in detection sensitivity while still achieving species-level specificity. Different FGAs have been reported to be capable of differentiating sequences with less than 87–94% sequence identity using oligonucleotide probes (50 or 70-mers). This is slightly higher than the 80–85% reported for a PCR amplicon FGA (Wu et al. 2001; Taroncher-Oldenburg et al. 2003; Rhee et al. 2004). Besides percent sequence identity, other factors including long stretches of a probe that perfectly match a nontarget sequence (Kane et al. 2000; Chen et al. 2006), the position of mismatches (Letowski et al. 2004), and the free energy of probe–target duplexes can also affect specificity (Li and Stormo 2001; Held et al. 2003). While it may be difficult to design probes that optimally match all these criteria using a sequential test and select process for each factor, it may be possible to use more relaxed criteria but yet produce specific probes if the factors are simultaneously considered during probe design (Liebich et al. 2006). This would increase the number of probes that can be designed for a given target and could be very important when designing probes from very similar sequence data (i.e., clone libraries) for which only a limited number of specific probes can be designed.

Several different software programs are currently available for the design of oligonucleotide microarray probes; however, many of these were developed for use with individual genomes and may design nonspecific probes when used with environmental sequence data (Li et al. 2005). A relatively new software program, called CommOligo, is now available that is capable of designing specific probes from orthologous gene sequences such as those produced by clone libraries (Li et al. 2005). CommOligo can design single or multiple unique probes for each sequence using multiple, user-specified criteria, including maximal sequence identity, maximal length of continuous perfectly matched nucleotides, free energy, self-binding, melting temperature, and GC content. Another feature of CommOligo is the ability to design group-specific probes for clusters of sequences whose sequence identities are too similar to each other to allow the design of unique probes for each sequence.

The specificity of microarray hybridizations can also be affected by increasing or decreasing the stringency of the hybridization conditions (e.g., temperature, formamide concentration, etc.) (Wu et al. 2001; Letowski et al. 2004; He et al. 2005b). For example, the specificity of one CGA could be increased from species-level to strain-level detection by increasing the hybridization temperature from 55 to 65 or 75°C (Wu et al. 2004). This may be useful for some applications in order to achieve more or less specificity depending on the research objective, but it also illustrates the need for careful use of microarrays in order to achieve the appropriate level of hybridization stringency for which the array was designed.

4.4 Quantitation

Several studies have indicated that microarrays can quantitatively detect target sequences in environmental samples (Wu et al. 2001, 2004; Rhee et al. 2004; Desantis et al. 2005). It has been demonstrated that short oligonucleotide arrays (e.g., POAs, WGAs) based on perfect match/mismatch probes pairs can quantitatively detect targets (r = 0.917) in the presence of background DNA (Desantis et al. 2005). However, POAs of this format are commonly used with PCR-amplified nucleic acids, which may complicate quantitative detection. Another study reported a good linear relationship (r^2 = 0.94) between a mixture of 11 different genes, varying in concentration from 1 pg to 1 ng, that were hybridized to a PCR amplicon FGA (Wu et al. 2001). Similarly, Rhee et al. (2004) found a linear relationship (r^2 = 0.95–0.99) for multiple genes, over a range of 75–1,000 ng genomic DNA, detected by a 50-mer FGA in the presence of background DNA. This study also demonstrated that FGA detection of mRNA was linear (r^2 = 0.96–0.99) over a range of 5.0×10^7–1.6×10^9 cells in the presence of background RNA. Linear detection (r^2 = 0.98), from 25 to 1,000 ng of DNA, has been reported for at least one CGA (Wu et al. 2004). Even with these examples of successful quantitative detection by microarrays, caution should be used when conducting experiments and interpreting data because of the potential variability introduced by multiple analytical steps including DNA extraction, amplification, labeling, hybridization, and analysis.

4.5 Data Analysis and Standardization

Another challenge for microarray analysis of uncultivated microbial communities is the availability and standardization of methods for data analysis. Many researchers currently adapt statistical methods that were initially developed for functional genomics of pure cultures. In some cases, it may be possible to use traditional statistical methods to analyze data, but these methods may not be adequate for analyzing the complex data sets that are frequently generated from environmental samples (Eyers et al. 2006). These problems will only be magnified as more comprehensive arrays are developed that will require more involved methods such as artificial neural networks in order to interpret the results (Pozhitkov et al. 2005). Furthermore, researchers commonly use different analytical methods for microarray construction, labeling, and hybridization, which can make it difficult to compare data between experiments and laboratories (Wilkes et al. 2007). The development and consistent use of universal standards for microarray experiments would help address these issues. One approach that appears promising is the use of labeled oligonucleotides or DNA fragments, spiked into the hybridization solution, as a control (Cho and Tiedje 2002; Dudley et al. 2002). This allows microarray results to be normalized on the basis of the signal intensity resulting from hybridization of these controls with corresponding control probes on the array.

5 Concluding Remarks

Comprehensive microarrays are now available for the study of uncultivated microorganisms. As the technology continues to progress rapidly, new advances will undoubtedly expand the potential of microarrays for microbial characterization even further. However, the true power of these arrays will begin to be realized only as they are used to investigate complex microbial communities and interactions in a variety of environments.

Acknowledgments The authors' efforts in preparing this chapter were supported by the United States Department of Energy Office of Science as part of its Genomics:GTL program through the Virtual Institute of Microbial Stress and Survival (VIMSS; http://vimss.lbl.gov) and Environmental Remediation Science Program of the Office of Biological and Environmental Research, Office of Science.

References

Adamczyk J, Hesselsoe M, Iversen N, Horn M, Lehner A, Nielsen PH, Schloter M, Roslev P, Wagner M (2003) The isotope array, a new tool that employs substrate-mediated labeling of rRNA for determination of microbial community structure and function. Appl Environ Microbiol 69:6875–6887

Barnett MJ, Toman CJ, Fisher RF, Long SR (2004) A dual-genome Symbiosis Chip for coordinate study of signal exchange and development in a prokaryote-host interaction. Proc Natl Acad Sci USA 101:16636–16641

Behr MA, Wilson MA, Gill WP, Salamon H, Schoolnik GK, Rane S, Small PM (1999) Comparative genomics of BCG vaccines by whole-genome DNA microarray. Science 284:1520–1523

Bent E, Yin B, Figueroa A, Ye J, Fu Q, Liu Z, McDonald V, Jeske D, Jiang T, Borneman J (2006) Development of a 9600-clone procedure for oligonucleotide fingerprinting of rRNA genes: utilization to identify soil bacterial rRNA genes that correlate in abundance with the development of avocado root rot. J Microbiol Methods 67:171–180

Berry AE, Chiocchini C, Selby T, Sosio M, Wellington EM (2003) Isolation of high molecular weight DNA from soil for cloning into BAC vectors. FEMS Microbiol Lett 223:15–20

Bjornstad A, Larsen BK, Skadsheim A, Jones MB, Andersen OK (2006) The potential of ecotoxicoproteomics in environmental monitoring: biomarker profiling in mussel plasma using ProteinChip array technology. J Toxicol Environ Health A 69:77–96

Bodrossy L, Stralis-Pavese N, Murrell JC, Radajewski S, Weilharter A, Sessitsch A (2003) Development and validation of a diagnostic microbial microarray for methanotrophs. Environ Microbiol 5:566–582

Bodrossy L, Stralis-Pavese N, Konrad-Koszler M, Weilharter A, Reichenauer TG, Schofer D, Sessitsch A (2006) mRNA-based parallel detection of active methanotroph populations by use of a diagnostic microarray. Appl Environ Microbiol 72:1672–1676

Botero LM, D'Imperio S, Burr M, McDermott TR, Young M, Hassett DJ (2005) Poly(A) polymerase modification and reverse transcriptase PCR amplification of environmental RNA. Appl Environ Microbiol 71:1267–1275

Brodie EL, Desantis TZ, Joyner DC, Baek SM, Larsen JT, Andersen GL, Hazen TC, Richardson PM, Herman DJ, Tokunaga TK, Wan JM, Firestone MK (2006) Application of a high-density oligonucleotide microarray approach to study bacterial population dynamics during uranium reduction and reoxidation. Appl Environ Microbiol 72:6288–6298

Brodie EL, DeSantis TZ, Parker JP, Zubietta IX, Piceno YM, Andersen GL (2007) Urban aerosols harbor diverse and dynamic bacterial populations. Proc Natl Acad Sci USA 104:299–304

Bunge J, Epstein SS, Peterson DG (2006) Comment on "Computational improvements reveal great bacterial diversity and high metal toxicity in soil". Science 313:918; author reply 918

Burgmann H, Widmer F, Sigler WV, Zeyer J (2003) mRNA extraction and reverse transcription-PCR protocol for detection of *nifH* gene expression by *Azotobacter vinelandii* in soil. Appl Environ Microbiol 69:1928–1935

Call DR, Bakko MK, Krug MJ, Roberts MC (2003) Identifying antimicrobial resistance genes with DNA microarrays. Antimicrob Agents Chemother 47:3290–3295

Castiglioni B, Rizzi E, Frosini A, Sivonen K, Rajaniemi P, Rantala A, Mugnai MA, Ventura S, Wilmotte A, Boutte C, Grubisic S, Balthasart P, Consolandi C, Bordoni R, Mezzelani A, Battaglia C, De Bellis G (2004) Development of a universal microarray based on the ligation detection reaction and 16S rRNA gene polymorphism to target diversity of cyanobacteria. Appl Environ Microbiol 70:7161–7172

Cebron A, Bodrossy L, Stralis-Pavese N, Singer AC, Thompson IP, Prosser JI, Murrell JC (2007) Nutrient amendments in soil DNA stable isotope probing experiments reduce the observed methanotroph diversity. Appl Environ Microbiol 73:798–807

Chandler DP, Jarrell AE (2004) Automated purification and suspension array detection of 16S rRNA from soil and sediment extracts by using tunable surface microparticles. Appl Environ Microbiol 70:2621–2631

Chandler DP, Newton GJ, Small JA, Daly DS (2003) Sequence versus structure for the direct detection of 16S rRNA on planar oligonucleotide microarrays. Appl Environ Microbiol 69:2950–2958

Chen YA, Chou CC, Lu X, Slate EH, Peck K, Xu W, Voit EO, Almeida JS (2006) A multivariate prediction model for microarray cross-hybridization. BMC Bioinformatics 7:101

Cho JC, Tiedje JM (2001) Bacterial species determination from DNA-DNA hybridization by using genome fragments and DNA microarrays. Appl Environ Microbiol 67:3677–3682

Cho JC, Tiedje JM (2002) Quantitative detection of microbial genes by using DNA microarrays. Appl Environ Microbiol 68:1425–1430

Cole JR, Chai B, Farris RJ, Wang Q, Kulam-Syed-Mohideen AS, McGarrell DM, Bandela AM, Cardenas E, Garrity GM, Tiedje JM (2007) The ribosomal database project (RDP-II): introducing myRDP space and quality controlled public data. Nucleic Acids Res 35:D169–D172

Denef VJ, Park J, Rodrigues JLM, Tsoi TV, Hashsham SA, Tiedje JM (2003) Validation of a more sensitive method for using spotted oligonucleotide DNA microarrays for functional genomics studies on bacterial communities. Environ Microbiol 5:933–943

Dennis P, Edwards EA, Liss SN, Fulthorpe R (2003) Monitoring gene expression in mixed microbial communities by using DNA microarrays. Appl Environ Microbiol 69:769–778

Desantis TZ, Stone CE, Murray SR, Moberg JP, Andersen GL (2005) Rapid quantification and taxonomic classification of environmental DNA from both prokaryotic and eukaryotic origins using a microarray. FEMS Microbiol Lett 245:271–278

Desantis TZ, Brodie EL, Moberg JP, Zubieta IX, Piceno YM, Andersen GL (2007) High-density universal 16S rRNA microarray analysis reveals broader diversity than typical clone library when sampling the environment. Microb Ecol 53:371–383

Dong Y, Glasner JD, Blattner FR, Triplett EW (2001) Genomic interspecies microarray hybridization: rapid discovery of three thousand genes in the maize endophyte, *Klebsiella pneumoniae* 342, by microarray hybridization with *Escherichia coli* K-12 open reading frames. Appl Environ Microbiol 67:1911–1921

Duburcq X, Olivier C, Malingue F, Desmet R, Bouzidi A, Zhou F, Auriault C, Gras-Masse H, Melnyk O (2004) Peptide-protein microarrays for the simultaneous detection of pathogen infections. Bioconjug Chem 15:307–316

Dudley AM, Aach J, Steffen MA, Church GM (2002) Measuring absolute expression with microarrays with a calibrated reference sample and an extended signal intensity range. Proc Natl Acad Sci USA 99:7554–7559

Dziejman M, Balon E, Boyd D, Fraser CM, Heidelberg JF, Mekalanos JJ (2002) Comparative genomic analysis of *Vibrio cholerae*: genes that correlate with cholera endemic and pandemic disease. Proc Natl Acad Sci U S A 99:1556–1561

El Fantroussi S, Urakawa H, Bernhard AE, Kelly JJ, Noble PA, Smidt H, Yershov GM, Stahl DA (2003) Direct profiling of environmental microbial populations by thermal dissociation analysis of native rRNAs hybridized to oligonucleotide microarrays. Appl Environ Microbiol 69:2377–2382

Eyers L, Smoot JC, Smoot LM, Bugli C, Urakawa H, McMurry Z, Siripong S, El-Fantroussi S, Lambert P, Agathos SN, Stahl DA (2006) Discrimination of shifts in a soil microbial community associated with TNT-contamination using a functional ANOVA of 16S rRNA hybridized to oligonucleotide microarrays. Environ Sci Technol 40:5867–5873

Gans J, Wolinsky M, Dunbar J (2005) Computational improvements reveal great bacterial diversity and high metal toxicity in soil. Science 309:1387–1390

Gao HC, Yang ZMK, Gentry TJ, Wu LY, Schadt CW, Zhou JZ (2007) Microarray-based analysis of microbial community RNAs by whole-community RNA amplification. Appl Environ Microbiol 73:563–571

Greene EA, Voordouw G (2003) Analysis of environmental microbial communities by reverse sample genome probing. J Microbiol Methods 53:211–219

Gregson BP, Millie DF, Cao C, Fahnenstiel GL, Pigg RJ, Fries DP (2006) Simplified enrichment and identification of environmental peptide toxins using antibody-capture surfaces with subsequent mass spectrometry detection. J Chromatogr A 1123:233–238

Grimm V, Ezaki S, Susa M, Knabbe C, Schmid RD, Bachmann TT (2004) Use of DNA microarrays for rapid genotyping of TEM beta-lactamases that confer resistance. J Clin Microbiol 42:3766–3774

Guschin D, Yershov G, Zaslavsky A, Gemmell A, Shick V, Proudnikov D, Arenkov P, Mirzabekov A (1997) Manual manufacturing of oligonucleotide, DNA, and protein microchips. Anal Biochem 250:203–211

Handelsman J (2004) Metagenomics: application of genomics to uncultured microorganisms. Microbiol Mol Biol Rev 68:669–685

He Z, Gentry TJ, Schadt CW, Wu L, Liebich J, Chong SC, Huang Z, Wu W, Gu B, Jardine P, Criddle C, Zhou J (2007) GeoChip: a comprehensive microarray for investigating biogeochemical, ecological and environmental processes. ISME J 1:67–77

He ZL, Wu LY, Fields MW, Zhou JZ (2005a) Use of microarrays with different probe sizes for monitoring gene expression. Appl Environ Microbiol 71:5154–5162

He ZL, Wu LY, Li XY, Fields MW, Zhou JZ (2005b) Empirical establishment of oligonucleotide probe design criteria. Appl Environ Microbiol 71:3753–3760

Held GA, Grinstein G, Tu Y (2003) Modeling of DNA microarray data by using physical properties of hybridization. Proc Natl Acad Sci USA 100:7575–7580

Hesse J, Jacak J, Kasper M, Regl G, Eichberger T, Winklmayr M, Aberger F, Sonnleitner M, Schlapak R, Howorka S, Muresan L, Frischauf AM, Schutz GJ (2006) RNA expression profiling at the single molecule level. Genome Res 16:1041–1045

Hurt RA, Qiu X, Wu L, Roh Y, Palumbo AV, Tiedje JM, Zhou J (2001) Simultaneous recovery of RNA and DNA from soils and sediments. Appl Environ Microbiol 67:4495–4503

Jenkins BD, Steward GF, Short SM, Ward BB, Zehr JP (2004) Fingerprinting diazotroph communities in the Chesapeake Bay by using a DNA macroarray. Appl Environ Microbiol 70:1767–1776

Kane MD, Jatkoe TA, Stumpf CR, Lu J, Thomas JD, Madore SJ (2000) Assessment of the sensitivity and specificity of oligonucleotide (50mer) microarrays. Nucleic Acids Res 28:4552–4557

Kim BC, Park JH, Gu MB (2004) Development of a DNA microarray chip for the identification of sludge bacteria using an unsequenced random genomic DNA hybridization method. Environ Sci Technol 38:6767–6774

Kingsley MT, Straub TA, Call DR, Daly DS, Wunschel SC, Chandler DP (2002) Fingerprinting closely related *Xanthomonas* pathovars with random nonamer oligonucleotide microarrays. Appl Environ Microbiol 68:6361–6370

Koizumi Y, Kelly JJ, Nakagawa T, Urakawa H, El-Fantroussi S, Al-Muzaini S, Fukui M, Urushigawa Y, Stahl DA (2002) Parallel characterization of anaerobic toluene- and ethylbenzene-degrading microbial consortia by PCR-denaturing gradient gel electrophoresis, RNA-DNA membrane hybridization, and DNA microarray technology. Appl Environ Microbiol 68:3215–3225

Korczak B, Frey J, Schrenzel J, Pluschke G, Pfister R, Ehricht R, Kuhnert P (2005) Use of diagnostic microarrays for determination of virulence gene patterns of Escherichia coli K1, a major cause of neonatal meningitis. J Clin Microbiol 43:1024–1031

Leigh MB, Pellizari VH, Uhlik O, Sutka R, Rodrigues J, Ostrom NE, Zhou J, Tiedje JM (2007) Biphenyl-utilizing bacteria and their functional genes in a pine root zone contaminated with polychlorinated biphenyls (PCBs). ISME J 1:134–148

Letowski J, Brousseau R, Masson L (2004) Designing better probes: effect of probe size, mismatch position and number on hybridization in DNA oligonucleotide microarrays. J Microbiol Meth 57:269–278

Li ES, Ng JK, Wu JH, Liu WT (2004) Evaluating single-base-pair discriminating capability of planar oligonucleotide microchips using a non-equilibrium dissociation approach. Environ Microbiol 6:1197–1202

Li F, Stormo GD (2001) Selection of optimal DNA oligos for gene expression arrays. Bioinformatics 17:1067–1076

Li XY, He ZL, Zhou JZ (2005) Selection of optimal oligonucleotide probes for microarrays using multiple criteria, global alignment and parameter estimation. Nucleic Acids Res 33:6114–6123

Liebich J, Schadt CW, Chong SC, He Z, Rhee SK, Zhou J (2006) Improvement of oligonucleotide probe design criteria for functional gene microarrays in environmental applications. Appl Environ Microbiol 72:1688–1691

Liu B, Bazan GC (2006) Synthesis of cationic conjugated polymers for use in label-free DNA microarrays. Nat Protoc 1:1698–1702

Liu RH, Yang J, Lenigk R, Bonanno J, Grodzinski P (2004) Self-contained, fully integrated biochip for sample preparation, polymerase chain reaction amplification, and DNA microarray detection. Anal Chem 76:1824–1831

Liu WT, Zhu L (2005) Environmental microbiology-on-a-chip and its future impacts. Trends Biotechnol 23:174–179

Liu WT, Mirzabekov AD, Stahl DA (2001) Optimization of an oligonucleotide microchip for microbial identification studies: a non-equilibrium dissociation approach. Environ Microbiol 3:619–629

Liu WT, Guo H, Wu JH (2007) Effects of target length on the hybridization efficiency and specificity of rRNA-based oligonucleotide microarrays. Appl Environ Microbiol 73:73–82

Loy A, Lehner A, Lee N, Adamczyk J, Meier H, Ernst J, Schleifer KH, Wagner M (2002) Oligonucleotide microarray for 16S rRNA gene-based detection of all recognized lineages of sulfate-reducing prokaryotes in the environment. Appl Environ Microbiol 68:5064–5081

Loy A, Kusel K, Lehner A, Drake HL, Wagner M (2004) Microarray and functional gene analyses of sulfate-reducing prokaryotes in low-sulfate, acidic fens reveal cooccurrence of recognized genera and novel lineages. Appl Environ Microbiol 70:6998–7009

Loy A, Schulz C, Lucker S, Schopfer-Wendels A, Stoecker K, Baranyi C, Lehner A, Wagner M (2005) 16S rRNA gene-based oligonucleotide microarray for environmental monitoring of the betaproteobacterial order "*Rhodocyclales*". Appl Environ Microbiol 71:1373–1386

Miao V, Davies J (2008) Metagenomics and antibiotic discovery from uncultivated bacteria. Microbil Monogr. doi: 10.1007/7171_2008_8

Moreno-Paz M, Parro V (2006) Amplification of low quantity bacterial RNA for microarray studies: time-course analysis of Leptospirillum ferrooxidans under nitrogen-fixing conditions. Environ Microbiol 8:1064–1073

Murray AE, Lies D, Li G, Nealson K, Zhou J, Tiedje JM (2001) DNA/DNA hybridization to microarrays reveals gene-specific differences between closely related microbial genomes. Proc Natl Acad Sci USA 98:9853–9858

Ochman H, Santos SR (2005) Exploring microbial microevolution with microarrays. Infect Genet Evol 5:103–108

Ochman H, Lerat E, Daubin V (2005) Examining bacterial species under the specter of gene transfer and exchange. Proc Natl Acad Sci USA 102 Suppl 1:6595–6599

Parro V, Moreno-Paz M (2003) Gene function analysis in environmental isolates: the nif regulon of the strict iron oxidizing bacterium *Leptospirillum ferrooxidans*. Proc Natl Acad Sci USA 100:7883–7888

Parro V, Moreno-Paz M, Gonzalez-Toril E (2007) Analysis of environmental transcriptomes by DNA microarrays. Environ Microbiol 9:453–464

Peplies J, Glockner FO, Amann R (2003) Optimization strategies for DNA microarray-based detection of bacteria with 16S rRNA-targeting oligonucleotide probes. Appl Environ Microbiol 69:1397–1407

Perreten V, Vorlet-Fawer L, Slickers P, Ehricht R, Kuhnert P, Frey J (2005) Microarray-based detection of 90 antibiotic resistance genes of gram-positive bacteria. J Clin Microbiol 43:2291–2302

Podar, Keller, Hugenholtz (2008) Single cell whole genome amplification of uncultivated organisms. Microbil Monogr. doi: 10.1007/7171_2008_10

Podar M, Abulencia CB, Walcher M, Hutchison D, Zengler K, Garcia JA, Holland T, Cotton D, Hauser L, Keller M (2007) Targeted access to the genomes of low-abundance organisms in complex microbial communities. Appl Environ Microbiol 73:3205–3214

Polz MF, Bertilsson S, Acinas SG, Hunt D (2003) A(r)Ray of hope in analysis of the function and diversity of microbial communities. Biol Bull 204:196–199

Pozhitkov A, Chernov B, Yershov G, Noble PA (2005) Evaluation of gel-pad oligonucleotide microarray technology by using artificial neural networks. Appl Environ Microbiol 71:8663–8676

Radajewski S, Ineson P, Parekh NR, Murrell JC (2000) Stable-isotope probing as a tool in microbial ecology. Nature 403:646–649

Rhee SK, Liu X, Wu L, Chong SC, Wan X, Zhou J (2004) Detection of genes involved in biodegradation and biotransformation in microbial communities by using 50-mer oligonucleotide microarrays. Appl Environ Microbiol 70:4303–4317

Ro HS, Jung SO, Kho BH, Hong HP, Lee JS, Shin YB, Kim MG, Chung BH (2005) Surface plasmon resonance imaging-based protein array chip system for monitoring a hexahistidine-tagged protein during expression and purification. Appl Environ Microbiol 71:1089–1092

Rudi K, Skulberg OM, Skulberg R, Jakobsen KS (2000) Application of sequence-specific labeled 16S rRNA gene oligonucleotide probes for genetic profiling of cyanobacterial abundance and diversity by array hybridization. Appl Environ Microbiol 66:4004–4011

Salama N, Guillemin K, McDaniel TK, Sherlock G, Tompkins L, Falkow S (2000) A whole-genome microarray reveals genetic diversity among *Helicobacter pylori* strains. Proc Natl Acad Sci USA 97:14668–14673

Saleh-Lakha S, Miller M, Campbell RG, Schneider K, Elahimanesh P, Hart MM, Trevors JT (2005) Microbial gene expression in soil: methods, applications and challenges. J Microbiol Methods 63:1–19

Sebat JL, Colwell FS, Crawford RL (2003) Metagenomic profiling: microarray analysis of an environmental genomic library. Appl Environ Microbiol 69:4927–4934

Sessitsch A, Gyamfi S, Stralis-Pavese N, Weilharter A, Pfeifer U (2002) RNA isolation from soil for bacterial community and functional analysis: evaluation of different extraction and soil conservation protocols. J Microbiol Methods 51:171–179

Steward GF, Jenkins BD, Ward BB, Zehr JP (2004) Development and testing of a DNA macroarray to assess nitrogenase (*nifH*) gene diversity. Appl Environ Microbiol 70:1455–1465

Stralis-Pavese N, Sessitsch A, Weilharter A, Reichenauer T, Riesing J, Csontos J, Murrell JC, Bodrossy L (2004) Optimization of diagnostic microarray for application in analysing landfill methanotroph communities under different plant covers. Environ Microbiol 6:347–363

Sunkara V, Hong BJ, Park JW (2007) Sensitivity enhancement of DNA microarray on nano-scale controlled surface by using a streptavidin-fluorophore conjugate. Biosens Bioelectron 22:1532–1537

Taroncher-Oldenburg G, Griner EM, Francis CA, Ward BB (2003) Oligonucleotide microarray for the study of functional gene diversity in the nitrogen cycle in the environment. Appl Environ Microbiol 69:1159–1171

Thompson M, Cheran LE, Zhang MQ, Chacko M, Huo H, Sadeghi S (2005) Label-free detection of nucleic acid and protein microarrays by scanning Kelvin nanoprobe. Biosens Bioelectron 20:1471–1481

Tiquia SM, Wu L, Chong SC, Passovets S, Xu D, Xu Y, Zhou J (2004) Evaluation of 50-mer oligonucleotide arrays for detecting microbial populations in environmental samples. Biotechniques 36:664–675

Torsvik V, Ovreas L (2002) Microbial diversity and function in soil: from genes to ecosystems. Curr Opin Microbiol 5:240–245

Tringe SG, von Mering C, Kobayashi A, Salamov AA, Chen K, Chang HW, Podar M, Short JM, Mathur EJ, Detter JC, Bork P, Hugenholtz P, Rubin EM (2005) Comparative metagenomics of microbial communities. Science 308:554–557

Tsai YL, Le JY, Olson BH (2003) Magnetic bead hybridization to detect enterotoxigenic *Escherichia coli* strains associated with cattle in environmental water sources. Can J Microbiol 49:391–398

Tyson GW, Chapman J, Hugenholtz P, Allen EE, Ram RJ, Richardson PM, Solovyev VV, Rubin EM, Rokhsar DS, Banfield JF (2004) Community structure and metabolism through reconstruction of microbial genomes from the environment. Nature 428:37–43

Urakawa H, El Fantroussi S, Smidt H, Smoot JC, Tribou EH, Kelly JJ, Noble PA, Stahl DA (2003) Optimization of single-base-pair mismatch discrimination in oligonucleotide microarrays. Appl Environ Microbiol 69:2848–2856

Van Nostrand JD, Khijniak TV, Gentry TJ, Novak MT, Sowder AG, Zhou JZ, Bertsch PM, Morris PJ (2007) Isolation and characterization of four gram-positive nickel-tolerant microorganisms from contaminated sediments. Microb Ecol 53:670–682

Venter JC, Remington K, Heidelberg JF, Halpern AL, Rusch D, Eisen JA, Wu D, Paulsen I, Nelson KE, Nelson W, Fouts DE, Levy S, Knap AH, Lomas MW, Nealson K, White O, Peterson J, Hoffman J, Parsons R, Baden-Tillson H, Pfannkoch C, Rogers YH, Smith HO (2004) Environmental genome shotgun sequencing of the Sargasso Sea. Science 304:66–74

Volkov I, Banavar JR, Maritan A (2006) Comment on "Computational improvements reveal great bacterial diversity and high metal toxicity in soil". Science 313:918; author reply 918

Wagner M, Nielsen PH, Loy A, Nielsen JL, Daims H (2006) Linking microbial community structure with function: fluorescence in situ hybridization-microautoradiography and isotope arrays. Curr Opin Biotechnol 17:83–91

Wagner M, Smidt H, Loy A, Zhou J (2007) Unravelling microbial communities with DNA-microarrays: challenges and future directions. Microb Ecol 53:498–506

Wan Y, Broschat SL, Call DR (2007) Validation of mixed-genome microarrays as a method for genetic discrimination. Appl Environ Microbiol 73:1425–1432

Wilkes T, Laux H, Foy CA (2007) Microarray data quality – review of current developments. Omics 11:1–13

Wilson KH, Wilson WJ, Radosevich JL, DeSantis TZ, Viswanathan VS, Kuczmarski TA, Andersen GL (2002) High-density microarray of small-subunit ribosomal DNA probes. Appl Environ Microbiol 68:2535–2541

Wu LY, Thompson DK, Li GS, Hurt RA, Tiedje JM, Zhou JZ (2001) Development and evaluation of functional gene arrays for detection of selected genes in the environment. Appl Environ Microbiol 67:5780–5790

Wu LY, Thompson DK, Liu XD, Fields MW, Bagwell CE, Tiedje JM, Zhou JZ (2004) Development and evaluation of microarray-based whole-genome hybridization for detection of microorganisms within the context of environmental applications. Environ Sci Technol 38:6775–6782

Wu LY, Liu X, Schadt CW, Zhou JZ (2006) Microarray-based analysis of subnanogram quantities of microbial community DNAs by using whole-community genome amplification. Appl Environ Microbiol 72:4931–4941

Yergeau E, Kang S, He Z, Zhou J, Kowalchuk GA (2007) Functional microarray analysis of nitrogen and carbon cycling genes across an Antarctic latitudinal transect. ISME J 1:163–179

Yilmaz LS, Noguera DR (2004) Mechanistic approach to the problem of hybridization efficiency in fluorescent in situ hybridization. Appl Environ Microbiol 70:7126–7139

Yilmaz LS, Okten HE, Noguera DR (2006) Making all parts of the 16S rRNA of Escherichia coli accessible in situ to single DNA oligonucleotides. Appl Environ Microbiol 72:733–744

Zhou X, Zhou J (2004) Improving the signal sensitivity and photostability of DNA hybridizations on microarrays by using dye-doped core-shell silica nanoparticles. Anal Chem 76:5302–5312

Zhou X, Wu L, Zhou J (2004) Fabrication of DNA microarrays on nanoengineered polymeric ultrathin film prepared by self-assembly of polyelectrolyte multilayers. Langmuir 20:8877–8885

Zhou J (2003) Microarrays for bacterial detection and microbial community analysis. Curr Opin Microbiol 6:288–294

The Seabed as Natural Laboratory: Lessons From Uncultivated Methanotrophs

Antje Boetius (✉), Thomas Holler, Katrin Knittel, Janine Felden, and Frank Wenzhöfer

Contents

Abstract The anaerobic oxidation of methane (AOM) by archaeal methanotrophs (ANME) functions as a major sink in oceanic methane geochemistry, and is a key biogeochemial process in the anoxic seabed. Unfortunately, demonstration of the biochemical pathway of AOM has not been possible because of the lack of pure cultures of ANME and their partner sulfate-reducing bacteria. The main reason for failing to isolate these microorganisms by cultivation is their slow growth, which is most likely a consequence of the low energy yield of the AOM reaction. This chapter discusses how in situ biogeochemical and microbiological observations of natural seabed communities and in vitro enrichments contribute to understanding of the ecology and physiology of these "uncultivables". Successful in vitro enrichment strategies include selecting seabed inoculates with abundant ANME populations, increasing the availability of dissolved methane and sulfate by flow through

A. Boetius, T. Holler, K. Knittel, J. Felden, and F. Wenzhöfer
Max Planck Institute for Marine Microbiology, Celsiusstr.1, 28359 Bremen, Germany

A. Boetius
Jacobs University Bremen, Campusring 1, 28759 Bremen, Germany
e-mail: aboetius@mpi-bremen.de

Microbiol Monogr, doi:10.1007/7171_2008_15

reactors and hydrostatic pressure, and maintaining the apparent temperature, pH, and salinity optima of AOM, the energy delivering process.

1 Introduction

Molecular analyses of DNA from seabed samples have revealed a vast diversity of microbial organisms, most of which have resisted cultivation. For microbes, the habitable zone in the seabed is huge and most of it is anoxic. It reaches down to 2–4 km below the seafloor, and ends only where temperatures exceed 120°C. At any depth, seabed sediments contain 10 to 1,000-fold more cells per volume than seawater, despite of their remoteness from the productive sea surface. Whitman et al. (1998) concluded that due to its tremendous size, the marine deep biosphere constitutes a "hidden majority" of all microbial cells on earth, equivalent to 50–80% of the global microbial biomass. From this it can be concluded that the average yet uncultivated microorganism on earth is a benthic marine anaerobe adapted to life at low energy.

Representing a major challenge for today's geomicrobiologists, these "uncultivables" include some of the globally most abundant and biogeochemically relevant microorganisms. Famous examples of anaerobic uncultivated microorganisms in the seabed are various sulfate-reducing *Deltaproteobacteria*, which inhabit cold marine sediments but escape isolation (Ravenschlag et al. 2000), benthic Crenarchaeota of unknown metabolism, which dominate 16S rRNA gene libraries of marine subsurface sediments (Biddle et al. 2006; Inagaki et al. 2006b), deep subsurface methanogens for which biogeochemical signatures are found, but no 16S rRNA gene sequences (Teske 2006), and the anaerobic methanotrophic Euryarchaeota (Hinrichs and Boetius 2002), which control methane emission from the ocean and are the focus of this chapter. Revealing the identity, functioning, and environmental role of such uncultivated microorganisms is paramount in understanding global element cycles and the diversity and genetic capacity of life on earth.

Many different approaches have been developed to overcome the problems associated with cultivation of marine anaerobes of the seabed, including carefully composed seawater media, laboratory techniques avoiding traces of oxygen penetrating the cultures, incubations at optimal pressure and temperature, and the use of sediment extracts, helper strains and various metabolites (Batzke et al. 2007). Targeted cultivation approaches are hampered by the uncertainties with regard to representative energy sources, biological and chemical interactions, and other in situ habitat characteristics. It is possible that unknown types of metabolisms prevail among the many uncultivated microorganisms of the seabed. The organic material buried at depths >1 m is thousands to millions of years old, and it remains unknown which fractions thereof can be used as electron donors. Assuming the absence of electron acceptors other than CO_2 and H_2O in deep sediments, it was proposed that methanogens and acetogens could be dominant groups in subsurface sediments, but these have rarely been detected at depth (Biddle et al. 2006; Inagaki et al. 2006a; Teske 2006).

Despite these uncertainties, the worst problem for cultivation of microorganisms could simply be the slow pace of microbial processes in the seabed. On the basis of the extremely low energy flux available to individual cells, theoretical generation times calculated for seabed microorganisms range from years to thousands of years (reviewed in Jørgensen and Boetius 2007). Generation times >2 months pose a problem to the isolation of microbial populations by competitive enrichment strategies, and often lead to a classification of the microbe as "uncultivable", simply because the time needed for cultivation exceeds the duration of PhD theses and project funding. Not much is known about the causes and controls of such extremely slow growth rates. Generally, anaerobes operate at much lower energy yields compared with aerobes. Some anaerobes are able to use energy for growth from reactions yielding less than -20 kJ mol^{-1}, which is often assumed as a threshold for microbial metabolism (Schink 1997; Jackson and McInerney 2002). The exact relation between energy availability, ATP synthesis, and growth remains unclear. Nauhaus et al. (2007) suggested that in microbial metabolisms at or below a certain threshold ΔG-value for ATP synthesis, substrate turnover, and growth could decouple, especially when a lot of activation energy is needed to channel the substrate into a metabolic pathway, as in anaerobic oxidation of methane (AOM). Consequently, most energy derived from such a reaction would be used for maintenance, i.e. replacement biosynthesis without net cell mass production, resulting in extremely low carbon assimilation efficiencies. However, in permanently anoxic habitats, where grazing pressure from eukaryotes is absent, the resulting slow growth may not be a problem for the survival of a microbial species, especially not if the microbe exploits a noncompetitive substrate.

In this regard, a special form of adaptation is found in syntrophic associations of microbes, which utilize the smallest energy yields known to be exploitable by any form of life (Jackson and McInerney 2002). In syntrophic associations, two species cooperate in a way that the degradation of a substrate by one species becomes thermodynamically possible via the removal of the product by another species. A variety of cooperations between sulfate-reducing *Deltaproteobacteria* and methanogens are known to function via an interspecies hydrogen transfer, which allows both organisms to grow from the smallest energy yields (Rabus et al. 2000). Syntrophic cooperation adds an extra problem to cultivation and isolation of microorganisms, as growth of one microorganism may depend on the partner.

This chapter focuses on uncultivated anaerobic methanotrophs (ANME) inhabiting the ocean floor and controlling methane fluxes from the seabed. Anaerobic methanotrophs are a presumably syntrophic consortium of uncultivated relatives of methanogenic archaea and sulfate-reducing bacteria (SRB). They have become an icon in the world of "uncultivables", highlighting how the lack of an axenic culture slows down progress in understanding of a climate-relevant process, despite the use of the most modern tools of molecular biology (Oremland et al. 2005).

2 The Discovery of Anaerobic Methanotrophs

In marine systems, anaerobic methanotrophs (ANME) appear to exclusively gain energy from the AOM with sulfate as the final electron acceptor according to the net reaction (1)

$$CH_4 + SO_4^{2-} \rightarrow HCO_3^- + HS^- + H_2O \quad (1)$$

This process was observed in marine sediments already in the seventies, but for decades the responsible microorganisms could not be identified. First evidence for AOM came from geochemical studies in marine sediments, which found that methane diffusing upwards from deeper sediment horizons disappeared in the same zone as sulfate, before any contact with oxygen (Martens and Berner 1974; Barnes and Goldberg 1976; Reeburgh 1976). Radioactive tracer incubations with ^{14}C-labeled methane and ^{35}S-labeled sulfate showed that methane oxidation coincided with increased sulfate reduction (Reeburgh 1980; Iversen and Jørgensen 1985). Zehnder and Brock (1979, 1980) investigated AOM in vitro by incubating a variety of methanogenic enrichments, and found evidence for degradation of methane at a small percentage of methane production. They proposed a cooperation between a methanogen responsible for the activation of methane together with a syntrophic partner acting as electron sink. On the basis of further investigations using coastal marine sediments, Hoehler et al. (1994) suggested that AOM could be performed by methanogenic archaea and SRB by interspecies hydrogen transfer. However, microbiologists still believed that because of its low energy yield, this reaction could not support life: The free energy change of AOM according to (1) under standard conditions at room temperature is only $\Delta G = -16$ kJ mol^{-1}, which would have to be shared by the two partners involved in AOM.

Intrigued by the growing geochemical evidence for AOM as a main process controlling methane emission from the seabed (reviewed in Reeburgh 2007), many microbiologists attempted to isolate a "reversible" methanogen using methane as the sole energy and carbon source for growth (Valentine and Reeburgh 2000) – unfortunately in vain. First, direct evidence for the existence of anaerobic methanotrophs came from Hinrichs et al. (1999) who extracted archaea-specific lipid biomarkers from a site on the continental slope off California (Eel River Basin) known to emit methane to the water column – a so called "cold seep". The archaeal lipids retrieved from Eel River Basin and from other sites such as Eastern Mediterranean mud volcanoes (Pancost et al. 2000) and gas hydrate bearing sediments from "Hydrate Ridge" (Elvert et al. 1999) were known from cultivated methanogens of the order *Methanosarcinales*, but here they were conspicuously depleted in the carbon isotope ^{13}C ($\delta < -100$ ‰ vs. PDB), indicating that methane was the carbon source for the organisms that synthesized the lipids. Interestingly, the rRNA gene library from Eel River Basin yielded new archaeal phylogenetic groups, one of which was a novel clade related to methanogens – and proposed by the authors to represent the anaerobic methanotrophs (Hinrichs et al. 1999).

The next step in the discovery of the ANME organisms was the visual identification of the methanotrophs via microscopy of cells hybridized with fluorochrome-labeled specific oligonucleotide probes (Boetius et al. 2000). Surprisingly, in sediments from Eel River Basin, Hydrate Ridge, and the Black Sea, conspicuous aggregates of archaea and SRB were highly abundant, representing >90% of the total microbial community. This finding supported the hypothesis of Hoehler et al. (1994) that methane could be used as substrate source via the cooperation of archaea able to activate methane, and sulfate-reducing bacteria providing an electron sink. The first in vitro experiment utilizing such sediments naturally enriched in ANME then successfully demonstrated that AOM coupled to sulfate reduction, according to the predicted net reaction (Nauhaus et al. 2002). The direct proof of anaerobic methanotrophy was provided by ion microprobe mass spectrometry confirming the extreme ^{13}C depletion of the aggregate biomass as predicted from the biomarker extractions (Orphan et al. 2002). Today it is known that anaerobic methanotrophic consortia are present wherever methane and sulfate cooccur, at a wide range of environmental conditions (Knittel et al. 2005). In fact, AOM consortia can form some of the densest and largest cell accumulations known to exist in nature (Michaelis et al. 2002). Three main clades of ANME (ANME-1, ANME-2, ANME-3) (Fig. 1) have been identified in ocean sediments and waters, and also in terrestrial habitats (Eller et al. 2005; Alain et al. 2006; Schubert et al. 2006). A fourth clade, (ANME-N) was found in a freshwater habitat and proposed to be involved in AOM utilizing nitrite as final electron acceptor (Raghoebarsing et al. 2006). The ANME-N associated nitrite-reducing bacteria, however, were recently shown to couple AOM without the involvement of the archaea (Ettwig et al. 2008).

It is interesting to note that AOM is associated with the slowest growth and lowest growth yield established so far for the microbial oxidation of an organic substrate with sulfate (Nauhaus et al. 2007). Growth yields of SRB utilizing various organic substrates mostly range between 4 and 18 g dry mass per mol sulfate (Widdel 1988; Rabus et al. 2000), depending on the "energy content" of the electron donor, but with methane it is less than 1 g (Nauhaus et al. 2007).

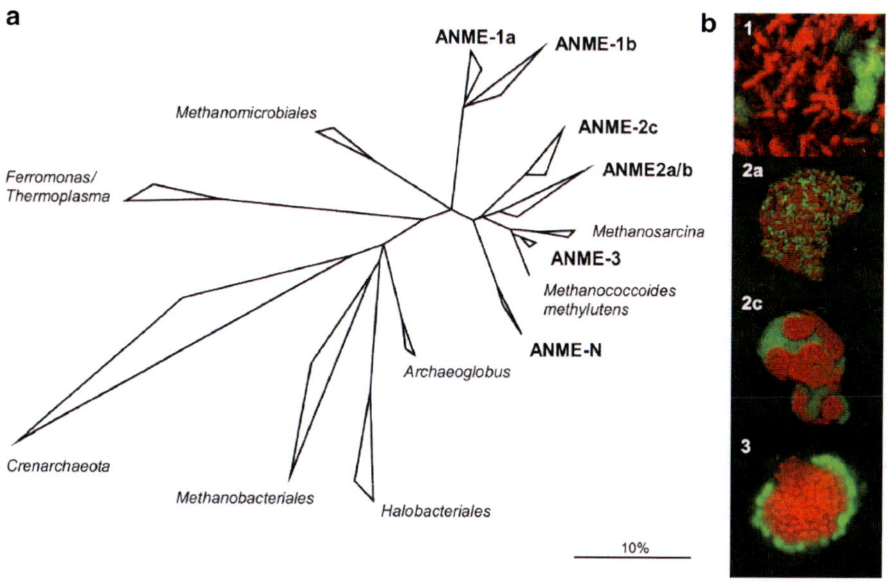

Fig. 1 Anaerobic methanotrophic archaea (ANME) (**a**) 16S rRNA gene based phylogeny of ANME-1, ANME-2, and ANME-3 clades; (**b**) micrographs of fluorescence in situ hybridization of ANME cells (*red*), from *top* to *bottom*: ANME-1, ANME-2a, ANME-2c, ANME-3. The partner SRB is shown in *green*. For specific probes see Table 1. (Source: project MUMM, K. Knittel, MPI)

This may explain why so far all attempts to enrich ANME populations "from scratch", on the basis of the idea that everything is everywhere (and just needs some methane), have failed. ANME sequences have rarely been retrieved from methane or sulfate-depleted sediments, so they cannot be grown from all types of seabed sample. Using naturally enriched communities, the process of AOM has been maintained in vitro in the laboratory and investigated with respect to substrate–product stoichiometry and dependence on pressure, temperature, and a variety of other environmental factors (Girguis et al. 2003, 2005; Nauhaus et al. 2002, 2005, 2007). Still, little is known about basic physiological parameters such as enzyme kinetics and growth yields of AOM under various conditions. Also the underlying biochemistry, the functional genes involved in AOM, and the nature of the interaction between both partners remain a matter of speculation.

3 Identification and Quantification of Anaerobic Methanotrophs in the Seabed

More than 1,500 ANME 16S rRNA gene sequences have been obtained from seabed samples around the world. These show that globally, the two most abundant and diverse phylogenetic groups of methane-oxidizing archaea are the ANME-1 and ANME-2 clades (Fig. 1). Both groups are usually associated with SRB of the SEEP-SRB I branch of the *Desulfosarcina/Desulfococcus* group (Knittel et al. 2003). A third phylogenetic group of archaea, (ANME-3) is associated with SRB of the *Desulfobulbus* group and has been mostly found at submarine mud volcanoes (Lösekann et al. 2007; Omoregie et al. in review) but also at other cold seeps (Orphan et al. 2001; Inagaki et al. 2004; Knittel et al. 2005) and recently in subsurface sediments (Parkes et al. 2007).

The ANME-2 and ANME-3 clades belong to the order *Methanosarcinales*, and the ANME-1 are more distantly related to the orders *Methanosarcinales* and Methanomicrobiales (Fig. 1, Knittel et al. 2005). The phylogenetic similarity between the three ANME clades is relatively low: the 16S rRNA gene sequences are only 75–92% related to each other, and the clusters of ANME-1 and ANME-2 are deeply branching. ANME-1 and ANME-2 as well as their sulfate-reducing partner bacteria share less than 92% 16S rRNA gene sequence similarity with any cultivated archaea and bacteria. The ANME-3 clade contains sequences, which are 95% similar to those of the *Methanococcoides* spp. Recent investigations have shown that the three ANME clades can also be distinguished on the basis of their *mcrA* genes (encoding the alpha subunit of the enzyme methyl coenzyme M reductase (MCR)), which appear to be phylogenetically congruent to the 16S rRNA gene (Hallam et al. 2003; Nunoura et al. 2006; Lösekann et al. 2007).

Today, most quantitative investigations of the distribution of ANME cells and their partner bacteria have relied on microscopical analyses using fluorescence in situ hybridization (FISH), with the exceptions of a few studies using slot blot hybridization (Treude et al. 2005), quantitative PCR of 16S rRNA genes (Girguis et al. 2003, 2005) or *mcrA* genes (Nunoura et al. 2006). Diverse forms of associations between ANME and their partner bacteria have been found (Fig. 1, Table 1), but

Table 1 Molecular identification of ANME consortia

Clade	ANME-1	ANME-2	ANME-3
16S rDNA Probe	ANME-1–350	ANME-2–538	ANME-3–1249
Sequence (5′-3′)	AGT TTT CGC GCC TGA TGC	GGC TAC CAC TCG GGC CGC	TCG GAG TAG GGA CCC ATT
Formamid conc.	40%	50%	20%
Reference	Boetius et al. 2000	Treude et al. 2005	Lösekann et al. 2007
Alpha subunit of methyl-coenzyme M-reductase	*mcr*A groups a-b	*mcr*A groups c-e	*mcr*A groups f
Reference	Hallam et al. 2003	Hallam et al. 2003	Lösekann et al. 2007
Typical morphology	Mat type (single cells)	ANME2a/b: mixed type ANME2c: shell type	Shell type (single cells)
Common SRB partner group	*Desulfococcus/ Desulfosarcina*	*Desulfococcus/ Desulfosarcina*	*Desulfobulbus*
Reference	Michaelis et al. 2002	Orphan et al. 2001	Lösekann et al. 2007
Lipid biomarker			
Archaeol	++	+	+
Hydroxyarchaeol	(+)	++	++
Crocetane	+	++	-
PMI	+ (0–5)	+ (0–4)	+ (4–5)
GDGT	++	(+)	–
Reference	Blumenberg et al. 2004	Nauhaus et al. 2007	Niemann et al. 2006

usually one phylogenetic group and one morphological type of consortium dominate a certain habitat (Knittel et al. 2005; Widdel et al. 2007). In the “shell-type” associations, an inner core of densely-packed archaeal cells is partially or fully surrounded by the bacterial partners. In the “mixed-type” consortia, archaeal and bacterial cells are more homogeneously distributed with neighboring archaeal and bacterial cells throughout the consortia. The “mat-type” is characterized by colonies of ANME and partners forming microbial mats. Single cells of ANME-1 and ANME-3 have also been found in situ as well as in in vitro enrichments, indicating that the physical association is not obligate, but the typical life mode of the anaerobic methanotrophs. Microscopical analysis of a growing ANME-2 population showed that population growth occurred as an increase in the number and size of the shell-type consortia (Nauhaus et al. 2007).

Another molecular method, which has been crucial in the investigation of the natural distribution of ANME populations, is the identification of specific lipid biomarkers and their stable carbon isotope signatures (Niemann and Elvert 2008). This method can integrate phylogenetic information with function (^{13}C signatures indicating methane assimilation) and can be used for comparative analyses of community biomass. All three ANME groups and their partner bacteria incorporate light (^{13}C-depleted) methane-derived carbon into their membrane lipids (Orphan et al. 2001; Niemann et al. 2006; Wegener et al. 2008). Using either naturally ^{13}C-depleted methane (Nauhaus et al. 2007), or ^{13}C-labeled methane as substrate

(Blumenberg et al. 2004; Wegener et al. 2008), experiments with environmental samples containing active populations of ANME have helped to identify typical membrane lipid profiles (Table 1). Crocetane is the only biomarker lipid specific for ANME-1 and ANME-2, i.e. not found in methanogenic cultures. The use of glycerol dibiphytanyl glycerol tetraether (GDGT) profiles as indicator of ANME distribution is difficult (Schouten et al. 2003), as the ANME GDGTs appear to overlap with those of benthic *Crenarchaeota*, which often share the same niche in the seabed (Knittel et al. 2005; Teske and Sorensen 2008). Because of the lack of cultures, the specificity of lipid biomarkers for ANME and the quantitative relation between cell abundances and lipid mass remains unknown.

An interesting question is the identity of yet uncultivated methanotrophs responsible for methane oxidation in low energy, diffusion driven sulfate-methane transition zones (SMTZ) of the deep biosphere. Recent investigations of subsurface SMTZ below 10 m sediment depth have not found evidence for the presence of ANME populations (Biddle et al. 2006; Inagaki et al. 2006b; Sorensen and Teske 2006). Hence, it has been proposed that other types of methanotrophs are responsible for AOM in such habitats (Biddle et al. 2006), e.g., members of the *Crenarchaeota*, which were found to be abundant and active in deep biosphere SMTZ (Sorensen and Teske 2006). Deep biosphere SMTZ are characterized by very low methane and sulfur fluxes of 1 $\mu mol\ m^{-2}\ yr^{-1}$ (Biddle et al. 2006) and AOM rates of $<1\ nmol\ cm^{-3}\ d^{-1}$ (Hoehler et al. 2000). Such low rates could be associated with very low densities of methanotrophs, of considerably less than 10^5 cells cm^{-3}. For biomarker extractions, the detection limit is as low as 1–10 ng of specific lipid cm^{-3} sediment. Hence, it is also possible that known ANME groups are present but not detected because of the very low population densities sustained at the respective low energy yields.

4 From In Situ Observations to In Vitro Enrichments

Studying the occurrence and distribution of natural ANME populations in the field has helped to choose seabed samples for further enrichment in the laboratory, for selecting enrichment conditions and for improving enrichment strategies. The following assumptions currently used to harvest and further enrich environmental ANME populations include:

1. High rates of methane oxidation concurrent with sulfate reduction and sulfide production indicate high densities of ANME populations.
2. Continuous high fluxes of both AOM substrates methane and sulfate at millimolar concentrations sustain high ANME biomasses of $>10^9$ cells cm^{-3}.
3. With regard to salinity, pH, and temperature, the apparent optima of AOM match the habitat settings within a relatively broad range. AOM (and probably growth) rates tend to be highest at 5–10°C degrees above the in situ temperature.
4. Oxygen is toxic for ANME, they do not occur in oxygen-penetrated sediments.

5. The full marine medium for sulfate-reducers, including ammonium as nitrogen source, meets the growth requirements of ANME.
6. The presence of the other partner is an essential factor in methane oxidation and growth.

4.1 Finding AOM Hotspots

The most common ANME habitat on earth is the so-called "sulfate-methane transition zone" (SMTZ) in the seabed (Thomsen et al. 2001; Niemann et al. 2006; Parkes et al. 2007). SMTZ are found in all seabed horizons where the diffusive transport of methane from below and sulfate from above leads to a zone of AOM (Reeburgh 2007 and references therein). Methane is completely consumed in the SMTZ, which may be found at decimeters to tens of meters below the seafloor (D'Hondt et al. 2004), depending on the depth of the methane production zone, the transport velocity of methane and sulfate, and their consumption rates. In diffusive seabed systems, the distribution of ANME is restricted to SMTZ. The SMTZ between 1 and 10 m sediment depth usually measure only a few centimeter to decimeter in vertical dimension, and are generally dominated by ANME 16S rRNA gene sequences. Population densities of ANME are low with $<10^6$ cells cm^{-3} (Niemann et al. 2005) and associated AOM rates are below 10 nmol cm^{-3} d^{-1} (<0.04 mol m^{-2} yr^{-1}). 16S rRNA gene libraries of ANME populations associated with the ubiquitous SMTZ zones are very similar to those of deep-water cold seeps. Most likely, the SMTZ-ANME represent the seed populations for cold seep communities. It remains unknown whether SMTZ-ANME possess special physiological adaptations to their energetically less favorable habitat compared with those at cold seeps.

Cold seeps form where tectonic or gravitational forces advect free gas, methane-rich porewater and/or muds upward into the sulfate-penetrated surface sediments and sometimes into the hydrosphere (Judd and Hovland 2007). Consequently, this higher energy availability leads to a natural enrichment of ANME populations and high AOM rates. The products of AOM, sulfide, and carbonate accumulate in the seabed, forming the typical features of cold seep ecosystems such as carbonate precipitates and high sulfide fluxes. The sulfide is oxidized chemically and microbially with oxygen, nitrate or iron close to the seafloor surface (de Beer et al. 2006). One of the most striking visual features of submarine cold seep ecosystems are mats of thiotrophic bacteria covering the seafloor. These mats are usually associated with high AOM rates and dense ANME communities inhabiting in the underlying sediments (Sahling et al. 2002; Treude et al. 2003; Joye et al. 2004; Niemann et al. 2006). Also, a variety of thiotrophic microbe-animal symbioses profit from AOM, such as siboglinid tubeworms, mytilid as well as vesicomyid bivalves (Sibuet and Olu 1998). Thiotrophic mats and invertebrate symbioses may enhance the growth of ANME populations by rapidly removing the toxic endproduct sulfide, and by replenishing sulfate in the sediments (Treude et al. 2003; Cordes et al. 2005;

Niemann et al. 2006). These specialized communities can thus be used as indicators to video-guided sampling of highly active sediments with high biomasses of ANME and their partner SRB (Fig. 2). Hence, the sedimentary environments of cold seeps appear to be ideal ANME habitats, providing a filter against oxygen and nitrate, and protecting the community against erosion by bottom water currents. In active cold seep sediments, AOM rates of 500–5,000 nmol CH_4 cm^{-3} d^{-1} (>2 mol m^{-2} yr^{-1}) are reached, associated with very dense ANME populations of >10^9 cells cm^{-3}. However, the active AOM zone at seeps can be very small, depending on sulfate depletion or purging by sulfate free subsurface fluids rising with high velocities (de Beer et al. 2006).

High methane fluxes are also found at hydrothermal vents of mid ocean ridges, but these ecosystems do not offer many niches for ANME populations. The seafloor of mid ocean ridges consists of basalts and lacks a sediment cover; hence, the habitats for ANME organisms are limited to small anoxic zones within vent chimneys (Schrenk et al. 2004). Furthermore, the fluids of most hydrothermal vents are sulfate free, and the seawater, which could provide sulfate, is oxygen rich – and hence toxic for ANME organisms (see Sect. 4.4). In contrast, sedimentary hydrothermal systems such as the Guaymas basin offer suitable habitats within the surface seafloor similar as the cold seep systems (Teske et al. 2002; Schouten et al. 2003).

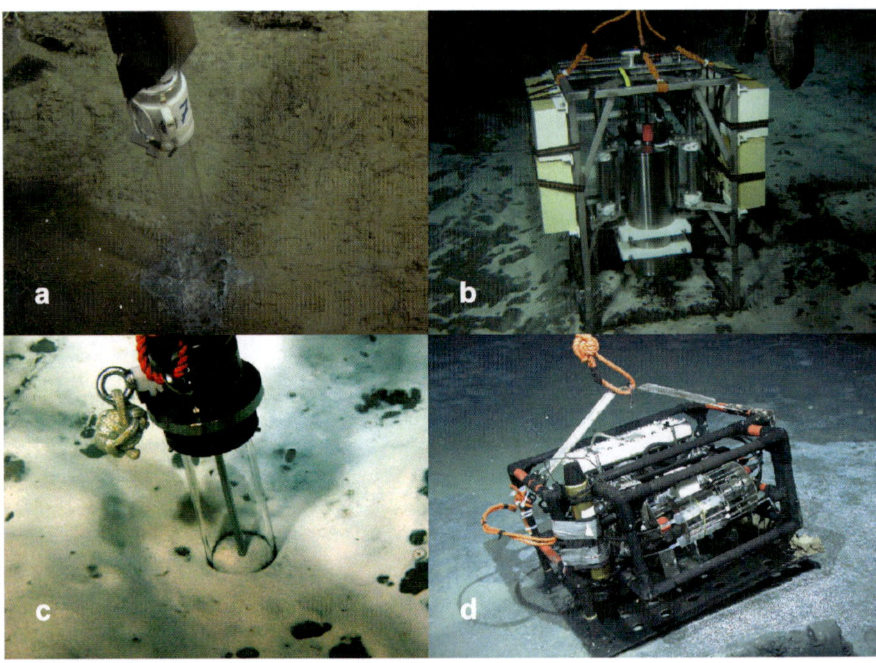

Fig. 2 Video guided coring and in situ measurement of methane fuelled sulfate reduction and sulfide production. (**a**) Sampling of a small mat of giant sulfide oxidizing bacteria, (**b**) in situ microsensor profiling in a microbial mat, (**c**) in situ measurements of sulfate reduction rates, (**d**) benthic chamber is placed by the ROV on microbial mat

Tools to retrieve ANME-inhabited environmental samples are TV-guided multiple corer, or push corer manipulated by ROV's or submersibles (Fig. 2). Undisturbed sediment samples can be used to analyze geochemical gradients in the seafloor (Fig. 3), to identify the peaks in methane oxidation, sulfate reduction, as well as sulfide and bicarbonate production, which coincide with the highest biomass of AOM consortia (Treude et al. 2003; Niemann et al. 2005; Lösekann et al. 2007). Most recently, we have used in situ microsensor profiling of sulfide concentrations at the seafloor, to identify the AOM hot spot before perturbation by recovery procedures (Niemann et al. 2006). In situ microsensor measurements allow characterizing the geochemical zonation of such habitats at a high vertical resolution (>50 μm) of different components (pH, T, H_2S, CO_2, O_2, Ca^{2+}) (Wenzhöfer et al. 2000, 2001; Wenzhöfer and Glud 2002; de Beer et al. 2006).

To quantify AOM or methane-fueled SR, intact seafloor sediments can be subsampled vertically with push cores for measurements of AOM and sulfate reduction (SR) using the whole core injection method (Jørgensen 1978; Treude et al. 2003). Radioactive tracers, i.e. $^{14}CH_4$ and $^{35}SO_4^{2-}$, are injected at centimeter resolution, and AOM and SR rates are calculated by multiplying the methane or sulfate reservoir with the percentage of $^{14}CO_2$ or $H_2^{35}S$ formed per time interval from the total tracer injected. Figure 3 shows the relation between AOM and SR as well as process-related chemical species in hydrate bearing sediments characterized by excess methane concentrations (Luff and Wallmann 2003). Active cold seep sediments show this peak directly below the surface, depending on the sulfate supply from

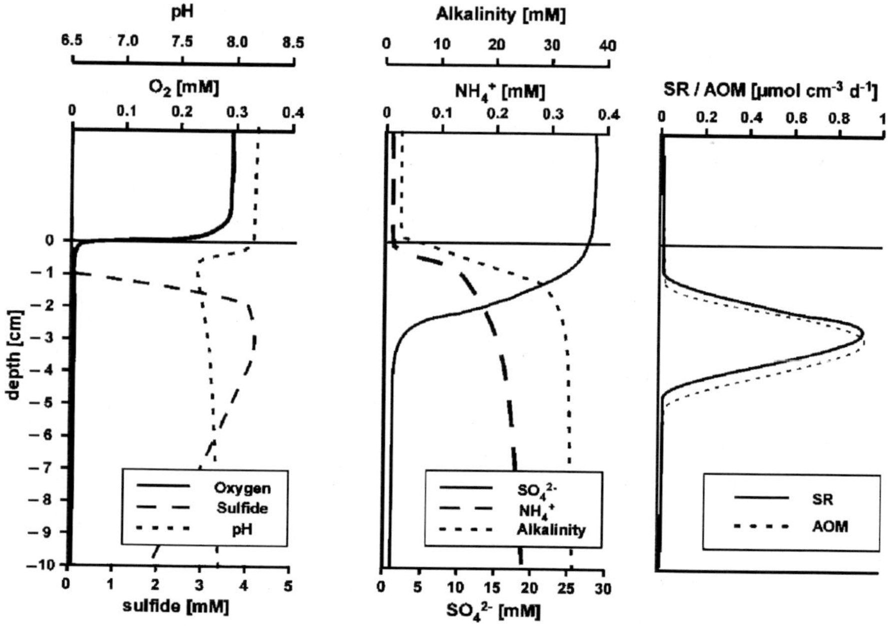

Fig. 3 Scheme of biogeochemical gradients in the AOM zone above hydrates

downward diffusing seawater. Identifying the usually very narrow zones of methanotrophic activity by their geochemical signatures is important for selecting the sediment horizons naturally enriched in AOM consortia.

4.2 Energetic Constraints: Supply of Methane and Sulfate, and Removal of End Products

In principle, cultivation of anaerobic methanotrophs should be possible if particular growth demands are known and used to design appropriate cultivation conditions in vitro. Most knowledge about the physiology and growth of ANME populations comes from incubations using methane seep sediments as inoculates, because they are naturally enriched in ANME biomass. To date, the most striking accumulations of ANME biomass have been found at highly active gas seeps on the anoxic Northwestern Black Sea shelf (Fig. 4, Michaelis et al. 2002; Knittel et al. 2005; Treude et al. 2007). Here, giant masses of microorganisms develop from subsurface aggregations into reef-forming microbial mats through which streams of gas bubbles emanate (Treude et al. 2005). These mats have been carefully harvested using submersible manipulators to obtain ANME-1 and ANME-2 biomass for physiology, genomics, and proteomics (Krüger et al. 2003; Meyerdierks et al. 2005). Interestingly, a high diversity of other bacteria and archaea has been found in the mats, apparently profiting from the methane-fueled biomass production, despite its slowness (Blumenberg et al. 2004; Reitner et al. 2005; Stadnitskaia et al. 2005). Three key factors appear to facilitate the development of such enormous methanotrophic biomasses – the continuous release of methane gas, the inward advection of sulfate by volume replacement, and the permanent anoxic conditions in the Black Sea water. However, due to the high microbial diversity and extremely low growth rate of the Black Sea microbial mats (Blumenberg et al. 2004; Treude et al. 2007), they represent a difficult inoculate for cultivation and isolation of ANME populations. Other types of well-investigated habitats hosting dense accumulations of ANME biomass are gas hydrate-laden surface seafloor of the NE Pacific margin (Eel River Basin and Hydrate Ridge; Orphan et al. 2001; Knittel et al. 2003; Treude et al. 2003) and the SE Atlantic margin (Pockmarks of the Congo fan; Olu-Le Roy et al. 2007), as well as mud volcano sediments of the Norwegian margin, Eastern Mediterranean, and Black Sea percolated with warm, methane-rich subsurface porewater at medium fluid flow velocities of 0.1–1 m yr^{-1} (Stadnitskaia et al. 2005; de Beer et al. 2006; Lösekann et al. 2007; Omoregie et al. 2008).

The solubility of methane depends on hydrostatic pressure, which increases with increasing water depth. Methane is essentially insoluble at atmospheric pressure (0.1 MPa) with a molar solubility of only 1.8 mM in seawater at 4°C. When methane builds up in the seabed and porewater gets oversaturated with methane, free gas bubbles form and may precipitate as gas hydrate at high pressures and cold temperatures. Gas hydrate is an ice-like solid formed by water crystals encaging large volumes of natural gas. The physicochemical stability zone of gas hydrates in the

Fig. 4 Methanotrophic mats of the Black Sea. (**a**) A methanotrophic microbial reef of the Crimea Area (Source: MARUM, M70/2 BIONIL), (**b**) Nodular mats on top of a reef structure, the nodules are about 10 cm in diameter (Source: JAGO, METROL project), (**c**) A nodule opened, consisting of *pinkish mat* below the black surface (Source: METROL project), (**d**) Scientists subsampling the mat for lipid biomarker and mineralogical analyses (Source: METROL project), (**e**) $^{14}CH_4$ incorporation into a mat section (top surface, bottom inside mat) visualized by beta-imaging. *Red* indicates high uptake. The FISH sections show that the most active methanotrophic zone in the mat coincides with the densest ANME population (*red*). Other zones of the mat show a high bacterial diversity (source: Treude et al. 2007)

seabed lies within water depths >500 m and temperatures <6°C. As the gas is in equilibrium with the surrounding porewater, the in situ methane concentrations reach tens to hundred millimolar in the vicinity of hydrates. However, the chemical energy stored in methane is only accessible to the anaerobic methanotrophs in the presence of sulfate. Hence, the most favorable natural enrichment conditions for ANME are to be expected around gas hydrates that lie within sulfate penetrated surface sediments (Boetius et al. 2000; Treude et al. 2003); or in the top surface sediments of active methane-emitting mud volcanoes (de Beer et al. 2006; Wallmann et al. 2006). In most methane-rich environments, sulfate is depleted rapidly, because the flux of methane from subsurface sediments often exceeds the flux of sulfate from the bottom waters. Accordingly, the highest densities of ANME populations can be expected where diffusion limitation is overcome, and sulfate gets advected into the seafloor e.g. in zones of gas ebullition, and below mats of motile Beggiatoa filaments (Sahling et al. 2002; Niemann et al. 2006) or burrowing animals such as the mud dwelling bivalve *Calyptogena* and siboglinid tubeworms, which may actively release sulfate through their roots (Sahling et al. 2002; Cordes et al. 2005; Niemann et al. 2006). Enrichment experiments with samples from such active cold seep sites have proven to be the most promising for in vitro enrichment of ANME populations (Girguis et al. 2003, 2005, Nauhaus et al. 2007, Holler et al. unpublished).

Obviously, one strategy for the enrichment of ANME is to simulate as closely as possible conditions prevailing in those marine habitats known to host high population densities. Using sediment from the Monterey Canyon seeps (East Pacific), Girguis and colleagues (2003, 2005) could stimulate an increase in 16S rRNA gene copies belonging to ANME groups with methane in a continuous medium flow system, and Wegener et al. (2008) achieved a significant uptake of ^{13}C-labeled methane into AOM consortia with another type of flow through reactor. The advantage of such flow through systems is that they provide a steady source of the electron donor methane and the electron acceptor sulfate, and at the same time remove the end products sulfide and bicarbonate. However, as long as such systems are operated under atmospheric pressures, the energy yield from AOM remains low (<25 kJ mol^{-1}) because of the insolubility of methane. A high-pressure flow-through bioreactor is under development, which is operated in a two-phase mode: high-pressure gas saturation of anoxic seawater medium is applied in the first phase, and long-term high-pressure flow-through incubation of ANME inoculate in the second phase for biomass generation (Deusner, unpublished data). However, semicontinuous slurry enrichments in tubes or bottles at intermediate methane pressures remain a helpful low-tech solution.

To provide methane to enrichment slurries kept in tubes or bottles, the gaseous methane can be injected through stoppers to the oxygen-free slurry headspace by syringes with hypodermic needles. At room temperature and ambient pressure, a volume of 24 ml of pure methane (101 kPa) is approximately 1 mmol. Nauhaus and colleagues (2002) could show that an increase of the methane partial pressure to 1.1 MPa (approximately 15 mM dissolved methane) clearly stimulated the rate of AOM in vitro by fivefold, indicating a very high saturation concentration for AOM. To apply such high partial pressures, a special incubation device has to be used (Fig. 5, Nauhaus et al. 2002). Thermodynamic calculations confirm that concentrations of methane of 60 mM will at least double the energy yield available for the methanotrophs, which could make a significant difference for the growth yield (Nauhaus et al. 2007).

However, even at high pressure, the principle techniques of isolation via selective enrichment of the targeted population, and outcompetition/dilution of others do not work with the anaerobic methanotrophs due to their slow growth (generation time of 2–7 months, Girguis et al. 2005; Nauhaus et al. 2007). Hence, to start enrichment cultures from seabed samples, it is relevant to obtain an estimate of the AOM activity and population density in the sediment sample to be used as inoculum. The dilution of the sediment with medium should be tailored to the AOM activity of the starter population in the original sediment – so that the consumption of methane and sulfate in the slurry can be monitored at reasonable time intervals. In AOM, the consumption of dissolved methane leads to a simultaneous production

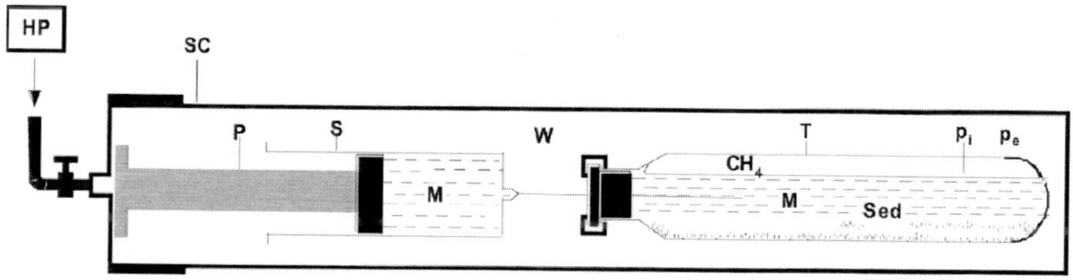

Fig. 5 Device for incubation of sediment (Sed) in seawater medium with methane of high partial pressure (high concentration). Slow increase of the surrounding water (W) pressure in the steel cylinder (SC) by means of a hydrostatic pump (HP) forces medium (M) from the syringe (S) by means of the piston (P) into the culture tube (T). The culture tube (total volume 20 ml) contains a methane headspace (CH_4). The volume of the medium in the syringe is slightly larger (approximately 1.2-fold) than the initial gas volume in the tube. This set-up allows complete dissolution of the gas phase and avoids implosion of the tube because the inner pressure (p_i) can equilibrate with any external hydrostatic pressure (p_e) applied (from Nauhaus et al 2002)

of sulfide at a molar ratio of 1:1 (Nauhaus et al. 2002). Hence, the simplest way of monitoring the ANME activity is measuring the increase in sulfide over time, if methane is the sole carbon source in the medium. For naturally enriched cold seep sediments with ANME densities of 10^9 cm^{-3}, inocula of > 20% volume are recommendable to obtain a significant increase in sulfide within days to weeks. Further dilutions should only be carried out when an exponential increase in methane-based sulfide production is observed as an indication for growth of the methanotrophic population.

It is important to overcome possible diffusion limitations of methane and sulfate in the enrichment cultures, for example by increasing the exchange area between headspace and medium, by using rotary tables, shaking of incubation vessels, etc. A slow but steady enrichment can be achieved via a semicontinuous culture technique, by letting the sediment settle in cultivation bottles, followed by a careful exchange of the medium under strict anoxic conditions, to replenish the substrates and get rid of the products, especially sulfide. Nauhaus et al. (2007) exchanged the medium when 10 mM concentrations of sulfide were reached, to avoid repression of AOM by end product accumulation. However, in nature, higher sulfide concentrations are often found.

4.3 *Temperature and pH Optima*

The temperature and pH optima for growth of specific ANME populations are not known. However, assuming that energy availability from AOM is the rate-limiting step in ANME growth, the apparent temperature and pH optimum of the energy delivering reaction (methane oxidation) may represent favorable growth conditions. For the free energy yield of the AOM reaction, the difference between the standard condition (25°C) and in situ conditions (e.g. 5°C) is negligible (Nauhaus et al. 2002). However, the effect of temperature on thermodynamic equilibrium concentrations of intermediates could be significant. In vitro experiments with a variety of naturally enriched seep sediments show that the apparent temperature optimum for AOM is usually found at 5–10°C above the in situ temperature (Table 2).

High resolution in situ pH measurements in marine sediments with high AOM activities indicate only small variations of pH between 7.7 and 7.9 in the AOM zone, and no substantial deviation of pH from the habitat appears to be caused by AOM (de Beer et al. 2006). In vitro experiments on the pH optimum of AOM with different ANME populations also showed a maximum within a range of pH 7–8, which is the typical pH range for marine sediments (Nauhaus et al. 2005). However, ANME populations have also been found in extreme environments: for example at temperatures of up to 95°C in the hydrothermal sediments of the Guaymas basin (Teske et al. 2002; Schouten et al. 2003), in the CO_2 vented sediments of the Yonaguni Knoll with in situ pH of probably as low as 4 (Inagaki et al. 2006b); in alkaline fluids of carbonate chimneys at Lost City at pH of 9–11 and temperatures of 55–75°C (Schrenk et al. 2004), and at high salt concentrations (Lloyd et al. 2006). Unfortunately, the physiological characteristics of such extremophilic anaerobic methanotrophs have not been studied yet.

So far, it remains unknown, which environmental factors select for which ANME groups and no direct relation between pH, temperature, and the presence of specific ANME groups has been found (Knittel et al. 2005; Krüger et al. 2005). It has been speculated that ANME-1 populations are better adapted to low energy environments (Blumenberg et al. 2004; but also see Girguis et al. 2005), and ANME-2 to high energy environments (Knittel et al. 2005), but usually a mixture of both or all three ANME clades are found in the environment, with one clade dominating the total archaeal biomass. However, it seems unlikely that specific

Table 2 Apparent temperature optima of AOM in vitro

Sampling site	Dominating ANME population	In situ temperature of the habitat	Optimum temperature of AOM	Reference
Haakon Mosby Mud Volcano	ANME-3/DBB	−1.5	4–8	Krüger et al. 2005
Hydrate Ridge	ANME-2/DSS	4	10–15	Nauhaus et al. 2005
Gulf of Mexico	ANME-2/DSS	6	16–20	Krüger et al. 2005
Black Sea	ANME-1/DSS	8	16–25	Nauhaus et al. 2005
Eastern Mediterranean	ANME-2/DSS	14	20	Holler et al. unpublished
Baltic Sea: Eckernförde Bay	ANME-2/DSS	5–20	20	Krüger et al. 2005; Treude et al. 2005
Baltic Sea: Kattegat	ANME-2/DSS, DBB	5–20	20–25	Krüger et al. 2005

environmental adaptations can be found on the level of the known clades, considering the relatively high phylogenetic distance of their members. Of the few investigations describing enrichments of ANME, it appears that the mixture of populations in the starter inoculum is also maintained in the growing biomass (Girguis et al. 2003, 2005, Blumenberg et al. 2004, Nauhaus et al. 2007). Hence, to obtain an enrichment of a specific group of ANME, it remains the best strategy to select an environmental sample dominated by this group.

4.4 Avoiding Oxygen

Both partner organisms – the ANME and the various sulfate-reducing bacteria associated with them – do not tolerate oxygen. Neither the ANME nor their sulfate-reducing partner bacteria have been detected in oxygen-penetrated sediments. The exact cause has not been investigated, but it is well known that methanogens are highly intolerant of oxygen because of the redox sensitivity of their enzyme cofactors (Jarrell 1985). This is most likely also true for their ANME relatives. Hence, determining the penetration depth of oxygen into the seafloor is important for identifying the ANME habitat. Most of the seafloor at continental margins consists of fine-grained clay sediments where oxygen penetration is restricted to a few centimeter to decimeter (Fig. 3). However, at fluid flow impacted cold seeps, oxygen penetration is limited to a few millimeter or less, because of the upward advection of anoxic porewaters. Furthermore, in organicrich settings, or at methane-laden seep sites, high amounts of sulfide are produced by sulfate reduction, causing rapid oxygen depletion, for example by mats of sulfideoxidizing bacteria (de Beer et al. 2006). At fluid flow impacted sites, a high spatial and temporal heterogeneity is observed, due to varying advective mass-transfer rates and due to bioturbation by macro and megafauna (Sahling et al. 2002; Niemann et al. 2006). With 1 and 2D measurements of the oxygen concentration in marine sediments by microsensor profiling and planar optode imaging (Fig. 2, Wenzhofer and Glud 2004; Glud et al. 2005), vertical as well as horizontal distribution patterns of oxygen in sediments can be obtained. These techniques allow continuous quantification of the O_2 distribution across the sediment–water interface at high spatial (~10–100 μm) and temporal resolution (seconds). Cold seep sediments show very high oxygen consumption, mostly driven by the methane, sulfide, and ammonium fluxes to the oxygenated surface sediments (Fig. 3).

After seafloor samples have been taken with cores or grabs, avoidance of oxygen penetration to AOM samples is a critical next step. In highly active sediments, the concentration of sulfide is elevated and represents a buffer against the toxic oxygen. Without problem, intact methane-rich sediment cores, which have not lost their bottom water, can be kept closed in the cold room at the environmental temperature before they are further processed. Subsampling in an anoxic glove box is not required for enrichment studies, but maybe essential for some types of porewater analyses. After filling the bottle with sediment, the headspace should be filled up with anoxic, sulfate-rich medium (Rabus et al. 2000) or purged with nitrogen, argon, or methane. Such samples can be stored for long, to be used for enrichments experiments even years after sampling (Nauhaus et al. 2002).

For the enrichment itself, it is essential to keep the sample anoxic during all handling steps. Removal of oxygen is most important in the preparation of the cultivation medium. Availability of an anoxic glove box is helpful, but with appropriate lab ware as for the cultivation of sulfate-reducing bacteria (Widdel and Bak 1992), all steps can be done at a normal laboratory bench. To prevent problems with oxygen leakage to the enrichment culture, addition of reducing agents (reductants) to the medium can be helpful at high dilution of the inoculate (e.g., <1:5 sediment to medium). However, in less diluted sediment slurries with active ANME populations, the sulfide production from methane oxidation will be high enough to prevent oxidation of the medium. Reductants that have often been used for the cultivation of methanogens or SRB are sulfide, dithionite, or cysteine (Widdel and Bak 1992). As a visual indicator of reducing conditions, resazurine may be added at a final concentration of 1 mg l^{-1}.

4.5 Other Growth Requirements

When methane – the most reduced form of organic carbon – is added as sole organic compound for cell synthesis to enrichment cultures, formation of the less reduced cell biomass, according to the bulk formula $C_4H_8O_2N$ (Harder and van Dijken 1975) requires a "biosynthetic net oxidation", which can be achieved by incorporation of CO_2 as the most oxidized form of carbon. Nauhaus et al. (2007) formulated the assimilation of methane as

$$17CH_4 + 15CO_2 + 8NH_4^+ \rightarrow 8C_4H_8O_2N + 14H_2O + 8H^+ \quad (2)$$

However, not much is known about the nitrogen requirement of the ANME populations. Nitrate and nitrite are absent from anoxic sediments, and regarding the low free energy yield of AOM, the uptake of ammonium seems more likely than nitrogen fixation. Anoxic marine sediments usually show relatively high concentrations of ammonium in the micromolar to millimolar range, hence sufficient reduced nitrogen should be available to support consortia growth. However, very recently nitrogen fixation has been suggested for methanotrophic consortia on the basis of the finding of the responsible genes in the metagenome and ^{15}N-enrichment in cellular nitrogen after $^{15}N_2$ incubation (Pernthaler et al. 2008).

Other salts, vitamins, and trace elements required for growth can be provided with the medium used for long-term enrichment. We are working with the full marine medium for SRB as described in Widdel et al. (2004), including the vitamin and trace element mixtures.

4.6 Inhibiting or Decoupling the Enrichment of ANME and Partner Bacteria

By addition of inhibitors, electron shuttles, or potential intermediates, it has been attempted to decouple the growth of one partner from the other (Nauhaus et al. 2002, 2005). The association of ANME with a sulfatereducing partner is most commonly interpreted as obligate syntrophic interaction in which the methanotrophic archaeon activates and metabolizes methane, leading to an intermediate that is scavenged as electron donor by the sulfate-reducing partner. However, the addition of a variety of electron-capturing shuttles such as phenazines and humic acids did not lead to decoupling of the ANME from the SRB (Nauhaus et al. 2005). Vice versa, in vitro "feeding" attempts with the conventional methanogenic substrates, H_2, formate, acetate, or methanol, did not succeed to decouple the SRB from the ANME, and no sulfide was formed by the partner SRB in the absence of methane (Nauhaus et al. 2002). The transfer of reducing equivalents from methane utilization into sulfate reduction could occur via an intermediate that is not a typical methanogenic growth substrate. It may occur by a transfer of reducing equivalents via an unknown organic compound, electron shuttles, or hydrogen shuttles other than free H_2, but so far, the nature of this presumed intermediate remains unknown. It has been proposed that permanent structures called "nanowires" can be established between microbial cells (Gorby et al. 2006), but this model is not favored by the finding that some ANME cells are not closely associated with bacterial partners (Orphan et al. 2002; Knittel et al. 2005).

5 Outlook: Other Isolation Techniques

If axenic cultures of the ANME cannot be obtained due to their slow growth and potentially obligate association with partner bacteria, an alternative strategy could be the mechanical isolation and enrichment of ANME cells. To obtain large fragments of DNA and proteins in high quality, it is important to isolate the microbial biomass from sediments and organic detritus. Important questions to the genome of both partners include for example if the ANME themselves may possess the genes for sulfate reduction, and what carbon and energy sources the sulfate reducers could utilize. Answering such questions requires the full genome sequence of ANME representatives and their partner bacteria. Metagenomic analyses of sediment communities have so far not arrived at closing genomes of the targeted organisms (Hallam et al. 2003; Meyerdierks et al. 2005; Pernthaler et al. 2008), and protein isolation and purification from sediments is extremely difficult and requires very dense biomasses (Krüger et al. 2003). First, successful application of an enrichment technique for ANME consortia included density gradient centrifugation with subsequent size selection; however, this did not result in pure biomass of a specific

ANME population (Orphan et al. 2001; Hallam et al. 2003; Pernthaler et al. 2008). New technologies such as single cell genome sequencing (Raghunathan et al. 2005), and single cell secondary ion mass spectrometry (SIMS) coupled to their phylogenetic identification (Kuypers and Jorgensen 2007) may allow a better understanding of AOM biochemistry, circumventing the need for an axenic cell culture. Methods that have recently been successfully combined with single cell or low biomass genome sequencing include laser forceps sorting of cells (Mitchell et al. 1993), fluorescence-activated cell sorting by flow cytometry (Stepanauskas and Sieracki 2007), microfluidics (Marcy et al. 2007), and magnetic bead sorting (Pernthaler et al. 2008). These, and other new technologies, combined with further physiological studies of selected environmental samples and in vitro enrichments will help to solve the enigmatic process of the AOM.

References

Alain K, Holler T, Musat F, Elvert M, Treude T, Kruger M (2006) Microbiological investigation of methane- and hydrocarbon-discharging mud volcanoes in the Carpathian Mountains, Romania. *Environ Microbiol* 8:574–590

Barnes RO, Goldberg ED (1976) Methane production and consumption in anoxic marine sediments. *Geology* 4:297–300

Batzke A, Engelen B, Sass H, Cypionka H (2007) Phylogenetic and physiological diversity of cultured deep-biosphere bacteria from Equatorial Pacific Ocean and Peru Margin sediments. *Geomicrobiol J* 24:261–273

Biddle JF et-al. (2006) Heterotrophic Archaea dominate sedimentary subsurface ecosystems off Peru. *Proc Natl Acad Sci USA* 103:3846–3851

Blumenberg M, Seifert R, Reitner J, Pape T, Michaelis W (2004) Membrane lipid patterns typify distinct anaerobic methanotrophic consortia. *Proc Natl Acad Sci USA* 101:11111–11116

Boetius A et-al. (2000) A marine microbial consortium apparently mediating anaerobic oxidation of methane. *Nature* 407:623–626

Cordes EE, Arthur MA, Shea K, Arvidson RS, Fisher CR (2005) Modeling the mutualistic interactions between tubeworms and microbial consortia. *Plos Biol* 3:497–506

D'Hondt S et-al. (2004) Distributions of microbial activities in deep subseafloor sediments. *Science* 306:2216–2221

de Beer D et-al. (2006) In situ fluxes and zonation of microbial activity in surface sediments of the Hakon Mosby Mud Volcano. *Limnol Oceanogr* 51:1315–1331

Eller G, Kanel LK, Krüger M (2005) Cooccurrence of aerobic and anaerobic methane oxidation in the water column of lake Plußsee. *Appl Environ Microbiol* 71:8925–8928

Elvert M, Suess E, Whiticar MJ (1999) Anaerobic methane oxidation associated with marine gas hydrates: superlight C-isotopes from saturated and unsaturated C-20 and C-25 irregular isoprenoids. *Naturwissenschaften* 86:295–300

Ettwing KF et-al. (2008) Denitrifying bacteria anaerobically oxidize methane in the absence of Archaea. Environ Microbiol doi:10.1111/j.1462-2920.2008.01724.x

Girguis PR, Cozen AE, DeLong EF (2005) Growth and population dynamics of anaerobic methane-oxidizing archaea and sulfate-reducing bacteria in a continuous-flow bioreactor. *Appl Environ Microbiol* 71:3725–3733

Girguis PR, Orphan VJ, Hallam SJ, DeLong EF (2003) Growth and methane oxidation rates of anaerobic methanotrophic archaea in a continuous-flow bioreactor. *Appl Environ Microbiol* 69:5472–5482

Glud RN, Wenzhofer F, Tengberg A, Middelboe M, Oguri K, Kitazato H (2005) Distribution of oxygen in surface sediments from central Sagami Bay, Japan: In situ measurements by microelectrodes and planar optodes. *Deep-Sea Res Pt I* 52:1974–1987

Gorby YA et-al. (2006) Electrically conductive bacterial nanowires produced by *Shewanella oneidensis* strain MR-1 and other microorganisms. *Proc Natl Acad Sci USA* 103:11358–11363

Hallam SJ, Girguis PR, Preston CM, Richardson PM, DeLong EF (2003) Identification of methyl coenzyme M reductase A (*mcrA*) genes associated with methane-oxidizing archaea. *Appl Environ Microbiol* 69:5483–5491

Harder W, van Dijken JP (1975) Theoretical considerations on the relation between energy production and growth of methane-utilizing bacteria. In: Schlegel HG, Gottschalk G, Pfennig N (eds) Microbial production and utilization of gases. Goltze, Göttingen, Germany, pp 403–418

Hinrichs K, Hayes J, Sylva S, Brewer P, DeLong EF (1999) Methane-consuming archaebacteria in marine sediments. *Nature* 398:802–805

Hinrichs KU, Boetius A (2002) The anaerobic oxidation of methane: new insights in microbial ecology and biogeochemistry. In: Wefer G, Billett D, Hebbeln D, Jorgensen BB, Schlüter M, van Weering T (eds) Ocean margin systems. Springer-Verlag, Heidelberg, pp 457–477

Hoehler TM, Alperin MJ, Albert DB, Martens CS (1994) Field and laboratory studies of methane oxidation in anoxic marine sediment: Evidence for a methanogenic-sulfate reducer consortium. *Glob Biogeochem Cycles* 8:451–463

Hoehler TM, Borowski WC, Alperin MJ, Rodriguez NM, Paull CK (2000) Model, stable isotope, and radiotracer characterization of anaerobic methane oxidation in gas hydrate-bearing sediments of the Blake Ridge. *Proc ODP Sci Res* 164:79–85

Inagaki F et-al. (2006a) Microbial community in a sediment-hosted CO2 lake of the southern Okinawa Trough hydrothermal system. *Proc Natl Acad Sci USA* 103:14164–14169

Inagaki F et-al. (2006b) Biogeographical distribution and diversity of microbes in methane hydrate-bearing deep marine sediments, on the Pacific Ocean Margin. *Proc Natl Acad Sci USA* 103:2815–2820

Inagaki F et-al. (2004) Characterization of C1-metabolizing prokaryotic communities in methane seep habitats at the Kuroshima Knoll, Southern Ryukyu Arc, by analyzing *pmoA*, *mmoX*, *mxaF*, *mcrA*, and 16S rRNA genes. *Appl Environ Microbiol* 70:7445–7455

Iversen N, Jørgensen BB (1985) Anaerobic methane oxidation rates at the sulfate-methane transition in marine sediments from Kattegat and Skagerrak (Denmark). *Limnol Oceanogr* 30:944–955

Jackson BE, McInerney MJ (2002) Anaerobic microbial metabolism can proceed close to thermodynamic limits. *Nature* 415:454–456

Jarrell KF (1985) Extreme oxygen sensitivity in methanogenic archaebacteria. *BioScience* 35:298–302

Jørgensen BB (1978) A comparison of methods for the quantification of bacterial sulfate reduction in coastal marine sediments. *I. Measurements with radiotracer techniques. Geomicrobiol J* 1:11–27

Jørgensen BB, Boetius A (2007) Feast and famine – microbial life in the deep-sea bed. *Nat Rev Microbiol* 5:770–781

Joye SB et-al. (2004) The anaerobic oxidation of methane and sulfate reduction in sediments from Gulf of Mexico cold seeps. *Chem Geol* 205:219–238

Judd AG, Hovland M (2007) Seabed fluid flow. Cambridge University Press, Cambridge

Knittel K et-al. (2003) Activity, distribution, and diversity of sulfate reducers and other bacteria in sediments above gas hydrate (Cascadia Margin, Oregon). *Geomicrobiol J* 20:269–294

Knittel K, Losekann T, Boetius A, Kort R, Amann R (2005) Diversity and distribution of methanotrophic archaea at cold seeps. *Appl Environ Microbiol* 71:467–479

Krüger M et-al. (2003) A conspicuous nickel protein in microbial mats that oxidize methane anaerobically. *Nature* 426:878–881

Krüger M, Treude T, Wolters H, Nauhaus K, Boetius A (2005) Microbial methane turnover in different marine habitats. *Palaegeogr Paleoclimatol Paleoecol* 227:6–17

Kuypers MMM, Jorgensen BB (2007) The future of single-cell environmental microbiology. *Environ Microbiol* 9:6–7

Lloyd KG, Lapham L, Teske A (2006) Anaerobic methane-oxidizing community of ANME-1b archaea in hypersaline Gulf of Mexico sediments. *Appl Environ Microbiol* 72:7218–7230

Lösekann T et-al. (2007) Diversity and abundance of aerobic and anaerobic methane oxidizers at the Haakon Mosby Mud Volcano, Barents Sea. *Appl Environ Microbiol* 73:3348–3362

Luff R, Wallmann K (2003) Fluid flow, methane fluxes, carbonate precipitation and biogeochemical turnover in gas hydrate-bearing sediments at Hydrate Ridge, Cascadia Margin: Numerical modeling and mass balances. *Geochim Cosmochim Acta* 67:3403–3421

Marcy Y et-al. (2007) Dissecting biological "dark matter" with single-cell genetic analysis of rare and uncultivated TM7 microbes from the human mouth. *Proc Natl Acad Sci USA* 104:11889–11894

Martens CS, Berner RA (1974) Methane production in the interstitial waters of sulfate-depleted marine sediments. *Science* 185:1167–1169

Meyerdierks A et-al. (2005) Insights into the genomes of archaea mediating the anaerobic oxidation of methane. *Environ Microbiol* 7:1937–1951

Michaelis W et-al. (2002) Microbial reefs in the Black Sea fueled by anaerobic oxidation of methane. *Science* 297:1013–1015

Mitchell JG, Weller R, Beconi M, Sell J, Holland J (1993) A practical optical trap for manipulating and isolating bacteria from complex microbial communities. *Microb Ecol* 25:113–119

Nauhaus K, Albrecht M, Elvert M, Boetius A, Widdel F (2007) In vitro cell growth of marine archaeal-bacterial consortia during anaerobic oxidation of methane with sulfate. *Environ Microbiol* 9:187–196

Nauhaus K, Boetius A, Kruger M, Widdel F (2002) In vitro demonstration of anaerobic oxidation of methane coupled to sulphate reduction in sediment from a marine gas hydrate area. *Environ Microbiol* 4:296–305

Nauhaus K, Treude T, Boetius A, Kruger M (2005) Environmental regulation of the anaerobic oxidation of methane: a comparison of ANME-I and ANME-II communities. *Environ Microbiol* 7:98–106

Niemann H et-al. (2005) Methane emission and consumption at a North Sea gas seep (Tommeliten area). *Biogeosciences* 2:335–351

Niemann H et-al. (2006) Novel microbial communities of the Haakon Mosby mud volcano and their role as a methane sink. *Nature* 443:854–858

Niemann H, Elvert M (2008) Diagnostic lipid biomarker and stable carbon isotope signatures of microbial communities mediating the anaerobic oxidation of methane with sulphate. Org Geochem doi:10.1016/j.orggeochem.2007.11.003

Nunoura T, Oida H, Toki T, Ashi J, Takai K, Horikoshi K (2006) Quantification of *mcrA* by quantitative fluorescent PCR in sediments from methane seep of the Nankai Trough. *FEMS Microbiol Ecol* 57:149–157

Olu-Le Roy K et-al. (2007) Cold-seep assemblages on a giant pockmark off West Africa: spatial patterns and environmental control. *Mar Ecol* 28:115–130

Omoregie EO et-al. (2008) Biogeochemistry and Community Composition of Iron- and Sulfur-Precipitating Microbial Mats at the Chefren Mud Volcano (Nile Deep Sea Fan, Eastern Mediterranean). *Appl Environ Microbiol* 74 (10):3198–3215

Oremland RS, Capone DG, Stolz JF, Fuhrman J (2005) Whither or wither geomicrobiology in the era of 'community metagenomics'. *Nat Rev Microbiol* 3:572–578

Orphan VJ, House CH, Hinrichs KU, McKeegan KD, DeLong EF (2001) Methane-consuming archaea revealed by directly coupled isotopic and phylogenetic analysis. *Science* 293:484–487

Orphan VJ, House CH, Hinrichs KU, McKeegan KD, DeLong EF (2002) Multiple archaeal groups mediate methane oxidation in anoxic cold seep sediments. *Proc Natl Acad Sci USA* 99:7663–7668

Pancost RD, Sinninghe Damste JS, de Lint S, van der Maarel MJEC, Gottschal JC, The Medinaut Shipboard Scientific Party (2000) Biomarker evidence for widespread anaerobic methane oxidation in Mediterranean sediments by a consortium of methanogenic archaea and bacteria. *Appl Environ Microbiol* 66:1126–1132

Parkes RJ et-al. (2007) Biogeochemistry and biodiversity of methane cycling in subsurface marine sediments (Skagerrak, Denmark). *Environ Microbiol* 9:1146–1161

Pernthaler A, Dekas AE, Brown CT, Goffredi SK, Embaye T, Orphan VJ (2008) Diverse syntrophic partnerships from deep-sea methane vents revealed by direct cell capture and metagenomics. *Proc Natl Acad Sci USA* 105:7052–7057

Rabus R, Hansen T, Widdel F (2000) Dissimilatory sulfate- and sulfur reducing procaryotes. In: Dworkin M, Falkow S, Rosenberg E, Schleifer K-H, Stackebrandt E (eds) The Prokaryotes, electronic edition. Fischer Verlag DOI: 10.1007/0–387–30742–7_22, New York

Raghoebarsing AA et-al. (2006) A microbial consortium couples anaerobic methane oxidation to denitrification. *Nature* 440:918–921

Raghunathan A, Ferguson HR, Bornarth CJ, Song WM, Driscoll M, Lasken RS (2005) Genomic DNA amplification from a single bacterium. *Appl Environ Microbiol* 71:3342–3347

Ravenschlag K, Sahm K, Knoblauch C, Jørgensen BB, Amann R (2000) Community structure, cellular rRNA content and activity of sulfate-reducing bacteria in marine Arctic sediments. *Appl Environ Microbiol* 66:3592–3602

Reeburgh WS (1976) Methane consumption in Cariaco Trench waters and sediments. *Earth Planet Sci Lett* 28:337–344

Reeburgh WS (1980) Anaerobic methane oxidation – rate depth distributions in Skan Bay sediments. *Earth Planet Sci Lett* 47:345–352

Reeburgh WS (2007) Oceanic methane biogeochemistry. *Chem Rev* 107:486–513

Reitner J, Reimer A, Schumann G, Thiel V, Peckmann J (2005) Methane-derived carbonate build-ups and associated microbial communities at cold seeps on the lower Crimean shelf (Black Sea). *Facies* 51:71–84

Sahling H, Rickert D, Lee RW, Linke P, Suess E (2002) Macrofaunal community structure and sulfide flux at gas hydrate deposits from the Cascadia convergent margin, NE Pacific. *Mar Ecol Prog Ser* 231:121–138

Schink B (1997) Energetics of syntrophic cooperation in methanogenic degradation. *Microbiol Mol Biol Rev* 61:262–280

Schouten S, Wakeham SG, Hopmans EC, Damste JSS (2003) Biogeochemical evidence that thermophilic archaea mediate the anaerobic oxidation of methane. *Appl Environ Microbiol* 69:1680–1686

Schrenk MO, Kelley DS, Bolton SA, Baross JA (2004) Low archaeal diversity linked to subseafloor geochemical processes at the Lost City Hydrothermal Field, Mid-Atlantic Ridge. *Environ Microbiol* 6:1086–1095

Schubert CJ et-al. (2006) Aerobic and anaerobic methanotrophs in the Black Sea water column. *Environ Microbiol* 8:1844–1856

Sibuet M, Olu K (1998) Biogeography, biodiversity and fluid dependence of deep-sea cold-seep communities at active and passive margins. *Deep-Sea Ros Pt II* 45:517

Sorensen KB, Teske A (2006) Stratified communities of active archaea in deep marine subsurface sediments. *Appl Environ Microbiol* 72:4596–4603

Stadnitskaia A et-al. (2005) Biomarker and 16S rDNA evidence for anaerobic oxidation of methane and related carbonate precipitation in deep-sea mud volcanoes of the Sorokin Trough, Black Sea. *Mar Geology* 217:67–96

Stepanauskas R, Sieracki ME (2007) Matching phylogeny and metabolism in the uncultured marine bacteria, one cell at a time. *Proc Natl Acad Sci USA* 104:9052–9057

Teske A et-al. (2002) Microbial diversity of hydrothermal sediments in the Guaymas Basin: evidence for anaerobic methanotrophic communities. *Appl Environ Microbiol* 68:1994–2007

Teske A, Sorensen KB (2008) Uncultured archaea in deep marine subsurface sediments: have we caught them all? *ISME J* 2:3–18

Teske AP (2006) Microbial communities of deep marine subsurface sediments: Molecular and cultivation surveys. *Geomicrobiol J* 23:357–368

Thomsen TR, Finster K, Ramsing NB (2001) Biogeochemical and molecular signatures of anaerobic methane oxidation in a marine sediment. *Appl Environ Microbiol* 67:1646–1656

Treude T, Boetius A, Knittel K, Wallmann K, Jorgensen BB (2003) Anaerobic oxidation of methane above gas hydrates at Hydrate Ridge, NE Pacific Ocean. *Mar Ecol Prog Ser* 264:1–14

Treude T, Knittel K, Blumenberg M, Seifert R, Boetius A (2005) Subsurface microbial methanotrophic mats in the Black Sea. *Appl Environ Microbiol* 71:6375–6378

Treude T, Orphan V, Knittel K, Gieseke A, House CH, Boetius A (2007) Consumption of methane and CO_2 by methanotrophic microbial mats from gas seeps of the anoxic Black Sea. *Appl Environ Microbiol* 73:2271–2283

Valentine DL, Reeburgh WS (2000) New perspectives on anaerobic methane oxidation. *Environ Microbiol* 2:477–484

Wallmann K, Drews M, Aloisi G, Bohrmann G (2006) Methane discharge into the Black Sea and the global ocean via fluid flow through submarine mud volcanoes. *Earth Planet Sci Lett* 248:545–560

Wegener G, Niemann H, Elvert M, Hinrichs K-U, Boetius A (2008) Assimilation of methane and inorganic carbon by microbial communities mediating the anaerobic oxidation of methane. *Environ Microbiol* 10:2287–2293

Wenzhöfer F et-al. (2001) Calcite dissolution driven by benthic mineralization in the deep-sea: in situ measurements of Ca^{2+}, pH, pCO_2 and O_2. *Geochim Cosmochim Acta* 65:2677–2690

Wenzhofer F, Glud RN (2004) Small-scale spatial and temporal variability in coastal benthic O_2 dynamics: effects of fauna activity. *Limnol Oceanogr* 49:1471–1481

Wenzhöfer F, Glud RN (2002) Benthic carbon mineralization in the Atlantic: a synthesis based on in situ data from the last decade. *Deep-Sea Res Pt I* 49:1255–1279

Wenzhöfer F, Holby O, Glud RN, Nielsen HK, Gundersen JK (2000) In situ microsensor studies of a shallow water hydrothermal vent at Milos, Greece. *Mar Chem* 69:43–54

Whitman WB, Coleman DC, Wiebe WJ (1998) Prokaryotes: The unseen majority. *Proc Natl Acad Sci USA* 95:6578–6583

Widdel F (1988) Microbiology and ecology of sulfate- and sulfur-reducing bacteria. In: Zehnder AJB (ed) Biology of anaerobic microorganisms. Wiley, New York, pp 469–585

Widdel F, Bak F (1992) Gram-negative mesophilic sulfate-reducing bacteria. In: Balows A, Trüper HG, Dworkin M, Harder W, Schleifer K-H (eds) The Prokaryotes, 2 edn. Springer-Verlag, New York, pp 3352–3378

Widdel F, Boetius A, Rabus R (2004) Anaerobic biodegradation of hydrocarbons including methane. In: Dworkin W, Rosenberg E, Schleifer KH, Stackebrandt E (eds) The Prokaryotes, electronic edition. Springer Science Online DOI: 10.1007/0–387–30742–7_33, New York

Widdel F, Musat F, Knittel K, Galushko A (2007) Anaerobic degradation of hydrocarbons with sulphate as electron acceptor. In: Barton LL, Hamilton WA (eds) Sulphate-reducing bacteria: environmental and engineered systems, 1st edn. Cambridge University Press, Cambridge, pp 265–303

Zehnder AJB, Brock TD (1979) Methane formation and methane oxidation by methanogenic bacteria. *J Bacteriol* 137:420–432

Zehnder AJB, Brock TD (1980) Anaerobic methane oxidation – occurrence and ecology. *Appl Environ Microbiol* 39:194–204

Single Cell Whole Genome Amplification of Uncultivated Organisms

Mircea Podar, Martin Keller, and Philip Hugenholtz (✉)

Contents

Abstract Whole genome amplification of single cells is emerging as a powerful technique for accessing the genomes of individual members of microbial communities without the complication of identifying the source of sequence data posed by shotgun sequencing of environmental samples (metagenomics). This method holds particular promise for the molecular unveiling of uncultivated organisms that comprise the bulk of the microbial diversity and functionality on our planet.

1 Introduction

Over the last two decades, it has become increasingly apparent that microbial diversity is massively undersampled by culture collections, and consequently coverage of the tree of life by reference genome sequences is also highly incomplete (Hugenholtz 2002; Rappe and Giovannoni 2003). This limitation greatly impedes our understanding

M. Podar and M. Keller
Oak Ridge National Laboratory, Biosciences Division, Oak Ridge, TN, USA

P. Hugenholtz
Microbial Ecology Program, DOE Joint Genome Institute, Walnut Creek, CA, USA
e-mail: phugenholtz@lbl.gov

Microbiol Monogr, doi:10.1007/7171_2008_10

of microbial evolution and ecology. Directed or fortuitous culturing efforts have managed to provide axenic or enriched cultures for some major phylogenetic gaps (Preston et al. 1996; Zhang et al. 2003; Zoetendal et al. 2003; Cho et al. 2004; Konneke et al. 2005), most of which immediately become the focus of genome sequencing projects (Liolios et al. 2006). To really accelerate genomic coverage of microbial diversity, however, methods not involving cultivation will need to be used. Metagenomics, the application of high-throughput sequencing to DNA extracted directly from environmental samples (Riesenfeld et al. 2004; Tringe and Rubin 2005), is a promising route to bypass the cultivation bottleneck, but has the drawback that multiple genomes are being sequenced simultaneously, often at very low coverage, which must be subsequently deconvoluted in silico. Therefore, metagenomic studies rarely result in complete or near-complete genomes, with the exception of communities highly enriched in one or more populations (Tyson et al. 2004; Garcia Martin et al. 2006; Strous et al. 2006). Whole genome amplification and sequencing of single cells has recently been demonstrated as a viable approach (Zhang et al. 2006), providing direct access to the full spectrum of microbial diversity without the multispecies complications of metagenomics. This opens the possibility for genomic characterization of any microbial member of a community that can be selected on the basis of taxonomic, morphological, or physiological criteria. As a result, acquiring genomic data for lineages encompassing the entire microbial tree of life appears feasible for the first time, regardless of cultivation success or abundance in the community for any specific type of organism.

Single cell genomics opens exciting opportunities to study a variety of aspects of microbial ecology and evolution. Understanding the degree of genomic variation between cells in natural populations not related by direct clonal descent will be important for defining the core genome vs. the accessory genes and ultimately getting better metrics for defining and distinguishing ecotypes, populations, and species or their equivalents. Since a coherent theoretical framework for microbial ecology is still lacking and important issues such as the definition or existence of bacterial species, and the extent and impact of horizontal gene transfer in microbial communities, genomic data from as many microbial lineages as possible will provide a better picture of the genetic and functional diversity in the microbial world and the evolutionary and environmental forces that shape genomes and communities.

In this chapter, we discuss the pros and cons of single cell genomics with particular reference to its recent application to members of candidate phylum TM7, a major lineage of the bacterial domain thus far lacking any pure-culture representatives and genome sequences.

2 Cell Isolation Methods

In most cases, a prerequisite to microbiological characterization of an organism has been the clonal expansion in the laboratory of a single cell, isolated from an environmental source. The resulting population, genetically uniform, serves for formal taxonomic description and further characterization of its physiology, genomics, and other biological aspects. Traditionally, obtaining that initial cell progenitor has involved techniques as simple as streaking on solid media or dilution to extinction. Such approaches are not broadly applicable to single cell genomics, as they offer little control over the types of organisms that are isolated and require active growth and formation of (micro)colonies or cultures. More discriminatory techniques such as micromanipulation with micro-capillaries or optical tweezers allow separation of specific cells based on morphology or, in some cases, certain physiological characteristics (e.g., autofluorescence) but have a low throughput and relatively restricted use. Such isolation techniques are the only viable options for subsequent genomic characterization if cells can grow at least to low densities on defined media in the laboratory.

2.1 Microdroplet Capture

A promising new approach to isolate microbial cells from the environment and test cultivation conditions in the laboratory involves encapsulation of individual cells into agarose microdroplets (Zengler et al. 2005). The microdroplets are then incubated under defined conditions that may allow formation of microcolonies, which can be detected microscopically or, using a higher throughput, by flow cytometry. If an appropriate selection can be designed (morphology, physiological property, or taxonomic staining by fluorescence in situ hybridization, see below), microdroplets containing microcolonies of interest can be isolated. When the cells are viable and dividing, a fraction of the microcolony can be further propagated in culture, while a reduced number of cells can be used directly for genomic characterization.

The above approaches, while suitable for cells that can be cultured in the laboratory, cannot be used for the isolation and genomic characterization of microbes that will not divide and establish a clonal population. The vast majority of microbes, while likely not intrinsically "unculturable" in the laboratory, fall into this category. Many major bacterial and archaeal lineages (phyla or divisions) still have no cultured representatives (Hugenholtz 2002) and all other phyla contain lower level taxa that are known solely on the basis of ribosomal RNA sequences obtained by culture-independent molecular surveys. Single cell genomics has great potential for characterizing such uncultured groups. In general, the most straightforward approach to identify and isolate single cells relies on the unique information available for those organisms, their ribosomal RNA sequences.

2.2 Micromanipulation

Antisense probes (oligonucleotides labeled with fluorescent small molecules) have been used for many years to detect and quantify specific microorganisms in environmental samples (Amann et al. 2001). The procedure, known as FISH (fluorescence in situ hybridization) involves the design of the antisense probe to target a region (usually 18–24 nt) of the ribosomal RNA sequence (usually the small subunit) that can confer the desired specificity (from species specificity to phylum or even domain) followed by chemical synthesis of the oligonucleotide and end labeling with a fluorophore that can be detected using UV, lasers, and specific filters. Software such as ARB (Ludwig et al. 2004) have probe design algorithms that facilitate the process and identify potential problems in terms of specificity, probe dimerization, and other artifacts that may occur. Studies on the in situ accessibility of ribosomal RNAs to FISH probes over the length of the rRNA molecule provide an additional guideline for probe design (Amann et al. 2001). Databases such as probeBase (Loy et al. 2007) catalog the sequences and characteristics of optimized probes, providing a valuable resource to the community. A large number of fluorophores emitting at different wavelengths are available commercially, which enable application of multiple probes with different specificities to be resolved on the basis of color combinations.

One of the limitations of FISH-directed cell isolation is that cells must be fixed and permeabilized and therefore are not viable and cannot potentially be used for establishing cultures. A variety of cell fixation protocols have been used, not all compatible with subsequent extraction and use of the nucleic acids for further characterizations. Formaldehyde, in particular, crosslinks DNA to protein (Speit et al. 2000), which severely impairs downstream applications. Milder fixation reagents such as ethanol have been used successfully and are compatible with enzymatic steps in DNA extraction protocols. Even so, fixation, hybridization conditions, probe specificity, and fluorescent signal strength are parameters that require optimization on a case-by-case basis. Once fluorescently labeled, the simplest approach to identify and isolate specific cells involves an epifluorescence microscope and a micromanipulator fitted with a microinjector. The main difficulty is contamination with nontarget cells and the tedious, low-throughput process. Nevertheless, this approach has been used successfully to isolate *Methanothermobacter thermoautotrophicus* as a test organism, followed by the isolation of a soil crenarchaeote (Kvist et al. 2007).

2.3 Fluorescence Activated Cell Sorting (FACS)

Another approach widely used to separate microorganism is flow cytometry, a fluorescence-based cell characterization technique that enables rapid analysis of entire cell populations on the basis of single cell characteristics (size, shape, cell count) as the individual cells are passed in front of an intense light source (laser or laser diode). When coupled with cell sorting based on specific fluorescent signal (FACS), a large number of cells from complex environmental samples can be rapidly analyzed and the ones containing the FISH probe can be isolated. A difficulty in using this approach is ensuring the sterility of the procedure and avoiding contamination of the separated cells with nontarget organisms and DNA. A secondary limitation is the low phylogenetic resolution of ribosomal RNAs due to their strong evolutionary conservation (Woese 1987); for example, 16S rRNA cannot typically resolve strains (genotypes). Therefore, if more than one cell is sorted, multiple genotypes may well be represented, complicating assembly of subsequent shotgun data. Using this approach, we have recently tested the feasibility of isolating from soil samples individual cells as well as small cell batches (5–100 cells) representing the uncultured TM7 phylum, after labeling with fluorescent oligonucleotides broadly specific for that group of bacteria (Podar et al. 2007). When sorting multiple cells, the different batches contained a varying degree of contamination with nontarget organisms or exogenous DNA; however, the TM7 cells appeared phylogenetically coherent (same species at least; Fig. 1) despite the broad specificity of the 16S rRNA-targeted FISH probe used. For example, a *Pseudomonas* cell was coisolated with four TM7 cells, resulting in significant contamination of the target TM7 genomes. The low frequency of the target cells in the soil community that served for that experiment (0.02%) indicates that the approach can be used to identify and isolate minor representatives of the community, "invisible" to other community genomic approaches. Obvious objectives for further improving the technique are finding ways to reduce the instances of contamination.

2.4 Microfluidics

Rapid developments in laser technology and nanotechnology have led to the miniaturization of flow cytometry and the development of an integrated microfabricated cell sorter using multilayer soft lithography (Fu et al. 2002). The integrated cell sorter is incorporated with various microfluidic functionalities, including peristaltic pumps, dampers, switch valves, and input and output wells, to perform cell sorting in a coordinated and automated fashion with extremely low fluidic volumes. Owing to the simple fabrication process and inexpensive materials, these devices can be disposable, thereby eliminating any cross-contamination from previous runs, which make them attractive for further downstream molecular methods requiring extreme sterility to avoid contaminating DNA. This benefit might compensate for the much lower sorting speed compared to conventional FACS machines; however, in practice it appears that removal of exogenous DNA entrained with the target cell is difficult even after repeated flushing of the microfluidic chamber perhaps because of the high surface tension in nanoliter volumes (Marcy et al. 2007). The cell sorting device can be directly combined with chemical or enzymatic reactions, such as cell lysis and DNA amplification. This approach has recently been used to characterize a different ecotype of TM7 phylum bacteria from the human subgingival microbiota ((Marcy et al. 2007); Fig. 1). Large rod-shaped cells, resembling oral TM7 previously characterized by FISH (Ouverney et al. 2003), were separated and lyzed in the microchamber, followed by DNA amplification in nanoliter scale volume to confirm identity and provide access to the genome of the isolated cell. Cross-contamination with exogenous DNA from other cell types present in the community still needs to be addressed; however, this approach allows selective isolation of viable cells based on morphology, which can then be analyzed genomically or transferred to culture media for propagation.

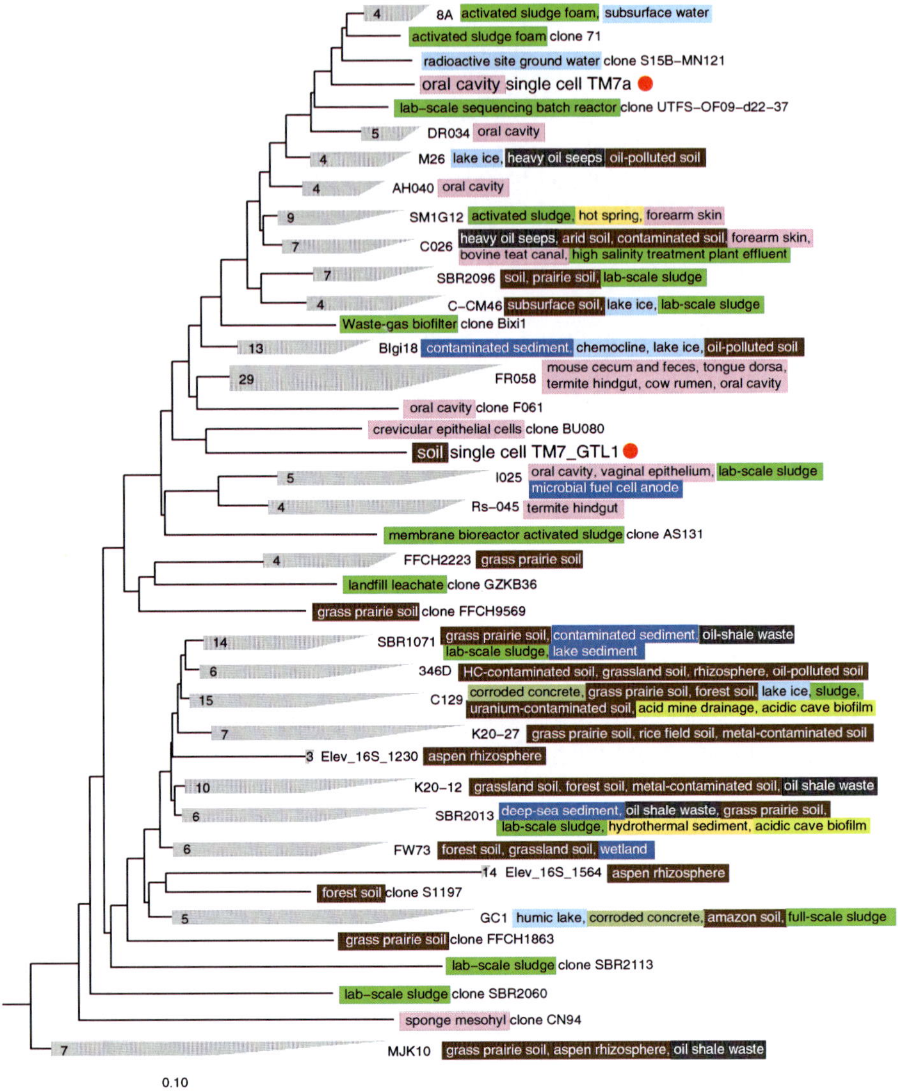

Fig. 1 A 16S rRNA gene tree of the bacterial phylum TM7 currently only comprising sequences obtained directly from the environment. Because of the large number of such environmental clone sequences, much of these data have been compressed into clusters named after the oldest clone representative, with the number indicating the number of clones in a given cluster. The habitats from which the clones were derived are listed and color-coded. The phylogenetic locations of amplified TM7 cells from soil (Podar et al. 2007) and the human oral cavity (Marcy et al. 2007) are indicated by red dots. The two ribotypes are ~9% divergent from each other. Data was obtained from the greengenes database (DeSantis et al. 2006)

3 DNA Isolation and Amplification from Single Cells

The amount of DNA present in a bacterial cell (~10^{-15} g) is presently well below the minimum amount necessary for cloning or direct sequencing. Amplification by factors of 10^6–10^9 is therefore required in order to generate nanogram to microgram quantities that can be used in standard genomic characterization protocols. Additional difficulties also arise because of handling of a single or a few chromosome copies. Very large DNA molecules are unstable when released from cells into solution and subject to single and double stranded breaks due to hydrodynamic shearing, nucleases, and chemical breakdown. While embedding of cells in gels, gentle lysis and handling, and inclusion of nuclease inhibitors can minimize such processes, they still occur but are averaged out in large DNA molecule populations that are obtained from bacterial cultures containing a large number of cells. When dealing with one or a few molecules, however, every single breakage event will result in the loss of information around that break point since the amplification enzymes may not be able to process them. Additional research will be required to understand the stability of whole chromosomes during cell lysis and enzymatic amplification and to determine whether genome sequencing to closure from a single molecule is feasible on a routine basis.

The current methods for DNA isolation from single bacterial cells include a cell treatment with lysozyme followed by alkaline denaturation (Zhang et al. 2006). Because the resulting DNA amount is very low, maintaining the volumes to a minimum is important in order to increase effective template concentration. Using conventional pipetting systems, reactions in several hundred nanoliters have been performed, which has resulted in amplification of single molecules to levels that allowed product visualization after gel electrophoresis (Hutchison and Venter 2006). Recent developments in the field of microfluidics allow, however, cell lysis and DNA amplification in much smaller volumes (tens of nanoliters) and therefore achieve an effective concentration of the template several orders of magnitude higher than using conventional pipetting (Hong et al. 2004). Such microfluidic chips that allow selection of target cell in addressable chambers followed by lysis and DNA amplification have been successfully used to generate sufficient DNA for partial genomic sequencing of uncultured TM7 bacteria from the human subgingival crevice (Marcy et al. 2007). The number of reports documenting the amplification of single molecules from uncultured organisms in small volumes is still small, and therefore factors such as efficient cell lysis and contamination minimization have to be investigated further.

Several different techniques have been published for random amplification across the whole genome, earlier work being done with Taq DNA polymerase (Hawkins et al. 2002). Multiple displacement amplification (MDA) is the approach that has been most successfully used for microbial genomic amplification and is based on using a highly processive phage polymerase with strand displacement capability, the phi29 DNA polymerase. Initial experiments using random hexamer oligonucleotides, circular phage genomic DNA, or plasmids as templates have resulted in 10,000-fold amplification by a multiple-primed rolling circle (RCA) mechanism (Dean et al. 2002). Sufficient DNA to perform cloning and sequencing was obtained in several hours at 30°C starting with a bacterial colony or a phage plaque. Commercial versions of the enzyme and optimized kits and protocols for MDA are available (e.g., GenomePlex by Sigma-Aldrich, REPLI-g by Qiagen, GenomiPhy by Amersham, RepliPHI by Epicentre), and are valuable for a wide range of medical, forensic, and microbiological applications in which the biological material is very limited. A recent study (Pinard et al. 2006) compared the available methods to evaluate the bias introduced during bacterial whole genome amplification, starting with 25 ng template DNA. The MDA reaction with phi29 polymerase vastly surpassed the amplification with Taq in both yield and absence of bias. However, it should be noted that the starting material was still much higher than that normally available when the reaction is performed using one or a few cells as DNA source.

There are several major difficulties when amplifying a single DNA molecule that have to be overcome before the technique can be robustly used to sequence whole microbial genomes. First, while phi29 is highly processive, it will still dissociate from the template on average after 10–20 kb (Blanco et al. 1989). Therefore, multiple primers hybridized to random positions along the chromosome can ensure broad amplification. However, because only a limited number of such primers can be positioned on the chromosome without affecting the reaction, a limited number of regions will be amplified from each individual chromosome, introducing a bias that is manifested when only one or a few initial templates are used, but averaged out in nanogram-amount template reactions. In the case of microbes for which clonal populations are available, a solution can be to perform multiple separate amplifications using single cells followed by pooling of the individual products prior to further characterization. When analyzing uncultured microbes from natural communities, however, populations are usually not clonal and therefore individual cells likely contain genomic differences. Reactions on multiple, separately collected cells could therefore allow the retrieval of a larger amount of "pangenomic" information for a given species, which can be valuable from ecological and evolutionary perspectives. However, if the goal is to assemble a complete genome, such heterogeneity may hinder genome finishing efforts, particularly if cells have been identified using 16S rRNA-targeting probes which may result in pooled cells with large amounts of genomic divergence due to the extreme conservation of the 16S rRNA molecule. Collecting cells that occur together in filaments or microcolonies will have a higher likelihood of being clonal copies. In analyzing an uncultured soil TM7 bacterium, we pooled five cells that were genetically related (identical small subunit ribosomal RNA sequence) and used them as template for a two-stage MDA in an effort to minimize the bias (Podar et al. 2007). On the basis of the assembled contigs, there did not appear to be a high frequency of single nucleotide polymorphisms; however, because the amplification was still biased and we retrieved only a relatively small fraction of the genome (~20%), evaluation of gene order and overall chromosomal conservation was difficult.

3.1 Chimerism

Another current limitation of MDA is the formation of chimeric structures, which result in fragmented genes, and difficulties in assembling large genomic contigs. Published data (Zhang et al. 2006) indicate that hyperbranched structures generated during the strand displacement amplification are resolved after ligation into vectors and cloning in *E.coli* as chimeric, rearranged sequences. To remove hyperbranched structures, Zhang and colleagues used an S1 nuclease to cut the junctions of branched DNA molecules, and constructed a 3-kb sequencing library. This library showed a significant improvement in comparison to the library constructed without S1 nuclease treatment (Zhang et al. 2006). However, the remaining chimeras still limited the quality of genome assemblies. An improved assembly with longer contigs was obtained by computationally splitting these chimeric sequences at their junction points based on a reference genome (Zhang et al. 2006). A new study (Lasken and Stockwell 2007) suggests, however, that most rearrangements are caused during the amplification step and that sequencing methods that would bypass the need for cloning may not necessarily avoid this problem. Computational tools and extensive sequence analysis appear effective in detecting and eliminating or flagging such rearrangements and, while not ideal, they will be a useful compromise until a biochemical solution is found.

A final difficulty in amplifying single bacterial chromosomes that should be mentioned is the high risk of contamination. Separating and handling single bacterial cells is technically challenging and there are numerous sources of potential contamination throughout the process, not to mention that when working with complex communities, avoiding isolation of more than one type of cell is very difficult. Not only are other cells potential contaminants but free DNA can be present in the sample, attached to cells, as well as in the reagents. For example, we have noted that *Delftia*, a common contaminant of laboratory reagents, appears to be present in some batches of MDA kits. Since any DNA present can be amplified, special precautions have to be taken to prevent introducing it into the reaction by ultraclean procedures and reagents. Even then, contamination usually occurs and can only be dealt with by recognizing and computationally removing the sequences that belong to the unwanted organism, using phylogenetics and other sequence characteristics (Marcy et al. 2007; Podar et al. 2007). For example, we used GC content binning to separate sequences that belong to a *Pseudomonas* contaminant from the bulk of the data representing the soil TM7 bacterium. The high GC content sequences contain genes with very high similarity values to other *Pseudomonas* genes and phylogenetically distant from organisms that appear to be more related to TM7 than Proteobacteria (Cyanobacteria, Chloroflexi, Firmicutes). As more refined methods to bin sequences from organisms based on sequence composition appear (McHardy et al. 2007), such computational approaches should improve the efficiency of filtering single cell genomic datasets for contaminant sequences.

In addition to single cell genome amplification, MDA has been used also to amplify environmental genomic DNA for shotgun sequencing. Low-biomass samples from highly contaminated soils yield DNA amounts that have limited use for direct, native analysis and screening. Using MDA, Abulencia and colleagues (Abulencia et al. 2006) recovered sufficient DNA from several low-biomass communities, which allowed diversity analysis as well as construction of genomic libraries for shotgun sequencing. While bias is difficult to control and evaluate in such cases, simply getting access to the genomes of uncultured organisms that inhabit such extreme environments should provide valuable information from both fundamental (ecological, physiological, and evolutionary) and practical (bioremediation) perspectives.

4 Sequencing and Genome Assembly

The strategy that has been applied almost exclusively so far for sequencing microbial genomes has been shotgun end sequencing of genomic DNA libraries using end chain terminator chemistry ("Sanger sequencing"). Over the last decade or so, tremendous improvements in this approach have been made, from long reads that approach 1,000 nucleotides, reduced cost associated with high-throughput platforms in specialized sequencing centers to efficient algorithms that allow assembly of the individual reads. Nevertheless, the time and effort to complete a genome project is still too long, primarily because of steps prior to the actual sequencing (e.g., genomic library construction, plasmid preps), and still too expensive ($8–$10k) to apply to more than a few organisms at a time in a standard laboratory or project. In the last several years a number of novel sequencing strategies have been explored that have the potential to trigger a new revolution in microbial genomic sequencing. One such approach, pyrosequencing, based on detection of the pyrophosphate released by the polymerase during copying of the sequencing template, was proposed two decades ago, but has only recently reached the stage of being applicable to genomic sequencing, primarily through an efficient parallel bead array–based technology developed by 454 Life Sciences (Ronaghi 2001). While the throughput of 454 sequencing is much higher than that of the traditional capillary Sanger sequencing (~100 Mbp vs. 3–4 Mbp per day per machine), the sequence reads are significantly shorter (~0.2 kbp vs. 1 kbp). The shorter reads present challenges for genome assembly, particularly in repetitive regions; however, a higher read depth coverage can address some of the limitations, and special assembly algorithms have been developed for the short sequence reads (Sundquist et al. 2007). An advantage of 454 sequencing is that genomic libraries do not need to be constructed, an important factor in sequencing genomes that have unusual sequence composition and are hard to clone into *E. coli*, or when the amount of starting DNA is limited such as in single cell genomics. Bypassing the cloning step in the latter case can avoid some of the recombinants that form in *E. coli* and are responsible for chimerism. However, it has been recently shown that a significant proportion of those chimeras form during amplification and not necessarily during cloning (Lasken and Stockwell 2007). Shorter reads will, in turn, make it more difficult to identify recombination points in the sequence. No direct comparisons have yet been made in sequencing genomes starting from one or a few cells using Sanger and pyrosequencing.

While not yet widely used as a prime approach for de novo sequencing of microbial genomes, 454 can be used efficiently in combination with Sanger sequencing (Goldberg et al. 2006) as well as for genome resequencing or sequencing of closely related strains. We have used both sequencing approaches separately in sequencing the two different members of the TM7 phylum. For the soil TM7, ~20,000 Sanger sequence reads were generated from a small insert genomic clone library, resulting in combined a contiguous sequence (contig) length of ~1.8 Mbp after initial assembly. Using an additional chimera detection strategy followed by annotation, taxonomic binning, and a second round of contig analysis, we eliminated sequences likely belonging to a *Pseudomonas* coisolate and further reduced the presence of recombinant chimeric sequences. The total length of the contigs assigned to the TM7 bacterium was ~0.6 Mbp, encoding approximately 600 genes. On the basis of functional gene categories accumulation curves as well as inferences based on universally present genes in bacterial genomes, the coverage of the soil TM7 was estimated to be of ~0.2×, which suggests a genome size of at least 3 Mbp. Using a sliding window analysis we detected sharp fluctuations in the coverage depth along the contigs, reflecting the known bias in amplifying the genomes of single cells as a result of the limited number of primer annealing events. However, analysis of the frequency of different types of functional gene categories represented in the soil TM7 genomic data suggested that the bias is not related to preferential amplification of certain types of genes.

Pyrosequencing using the 454 technology was applied in the case of three individual TM7 cells isolated from the human oral microbial community by microchip fluidics (Marcy et al. 2007). As is the case with other single cell separation techniques including flow cytometry, contamination is still a problem and manifested in the microfluidic chips separation by the presence of small amounts of hitchhiking DNA from *Leptotrichia*, a recognized member of the oral community. While no cloning was involved, genomic sequence analysis revealed, as with the soil TM7, chimeric sequences and bias in genome coverage. A different assembly strategy had to by used, specifically designed to handle the short reads generated during pyrosequencing, namely the 454 Newbler assembler and the Forge whole genome shotgun assembler (D. Platt, unpublished). Overall, from the cell that yielded the highest quality sequence, 1,474 genes on 288 contigs were obtained. The gene prediction software, Fgenesb (http://www.softberry.com/) was used to predict genes, and has been shown to be perform relatively well on simulated metagenomic datasets (Mavromatis et al. 2007), followed by analysis using the IMG/M database platform (Markowitz et al. 2008).

The two described projects aimed at genomic analysis of organisms from the TM7 phylum employed different samples and different experimental and computational strategies. While overall the two studies encompass the same concepts of single cell isolation, genome amplification, and sequencing, it will be important in future to do side-by-side comparisons of contamination likelihood, amplification efficiency, and the quality of the resulting DNA sequences when applying flow cytometry and chip microfluidics to the same environmental sample.

5 From Genomes to Biology

While traditionally selection of microorganisms for genome sequencing is based on prior knowledge of an organism's physiology, ecology, or other biological aspects, the TM7 phylum was targeted because of its novel phylogenetic position and lack of cultivated representatives. Not surprisingly, little is known about the numerous uncultured bacterial phyla highlighted by culture-independent surveys (Hugenholtz 2002); however, TM7 has received somewhat more attention, providing a small amount of physiological and ultrastructural data with which to compare genome data. TM7 bacteria are known to inhabit a wide range of environments, from soils, water, and activated sludge, to termite guts ((Hugenholtz et al. 1998), Fig. 1). They have also been found in the human oral cavity and positively correlated with mild periodontitis (Brinig et al. 2003; Kumar et al. 2003). There is microscopic evidence that members of the TM7 phylum are morphologically diverse, and large conspicuous sheathed TM7 filaments often have epiflora (Hugenholtz et al. 2001; Thomsen et al. 2002; Xia et al. 2008). Electron microscopy also revealed that the cell envelope of a sheathed filament TM7 morphotype from activated sludge is ultrastructurally indistinguishable from Gram-positive bacteria, raising the possibility that TM7 represents a third Gram-positive bacterial phylum (Hugenholtz et al. 2001). If correct, TM7 may be able to help shed light on the evolutionary origins of the Gram-positive cell envelope (i.e., monophyletic vs. paraphyletic origin). There has been a sustained effort to obtain a glimpse into the physiology of TM7 bacteria in the absence of pure cultures, revealing that bacteria belonging to the TM7 phylum are capable of protein hydrolysis (Xia et al. 2007) and can uptake a number of carbon substrates under aerobic and anaerobic conditions (Thomsen et al. 2002). TM7 bacteria from soil have been shown to form microcolonies on a membrane support (Ferrari et al. 2005), although formal species description and continuous cultivation have not yet been reported for any member of this phylum. Genomic information may in fact facilitate cultivation attempts as has been recently demonstrated for bacteria from an acidophilic community (Tyson et al. 2005).

As is usually the case when genomic information first becomes available for members of a microbial phylum, it provides an opportunity to analyze evolutionary relationships with other organisms using sequence data other than the traditional ribosomal RNA. While rRNA trees are remarkably consistent with genome trees (Wolf et al. 2002), the latter can provide greater resolution because they incorporate a larger amount of phyogenetically informative data. For example, on the basis of comparative analysis of concatenated conserved marker genes, the bacterial phylum Acidobacteria appears to be specifically related to the Deltaproteobacteria (Ciccarelli et al. 2006), a relationship that was not detected by 16S rRNA gene comparisons. Previous analyses of environmental rRNA sequences have shown no specific affiliation of TM7 to any other bacterial phylum; indeed, this was the basis for classifying TM7 as a candidate phylum in the first place (Hugenholtz et al. 1998). Using concatenated ribosomal protein sequences, we were able to show that TM7 is most likely a sister group of the Chloroflexi (green nonsulfur bacteria, Fig. 2). This may indicate that the Gram-positive cell envelope has arisen more than

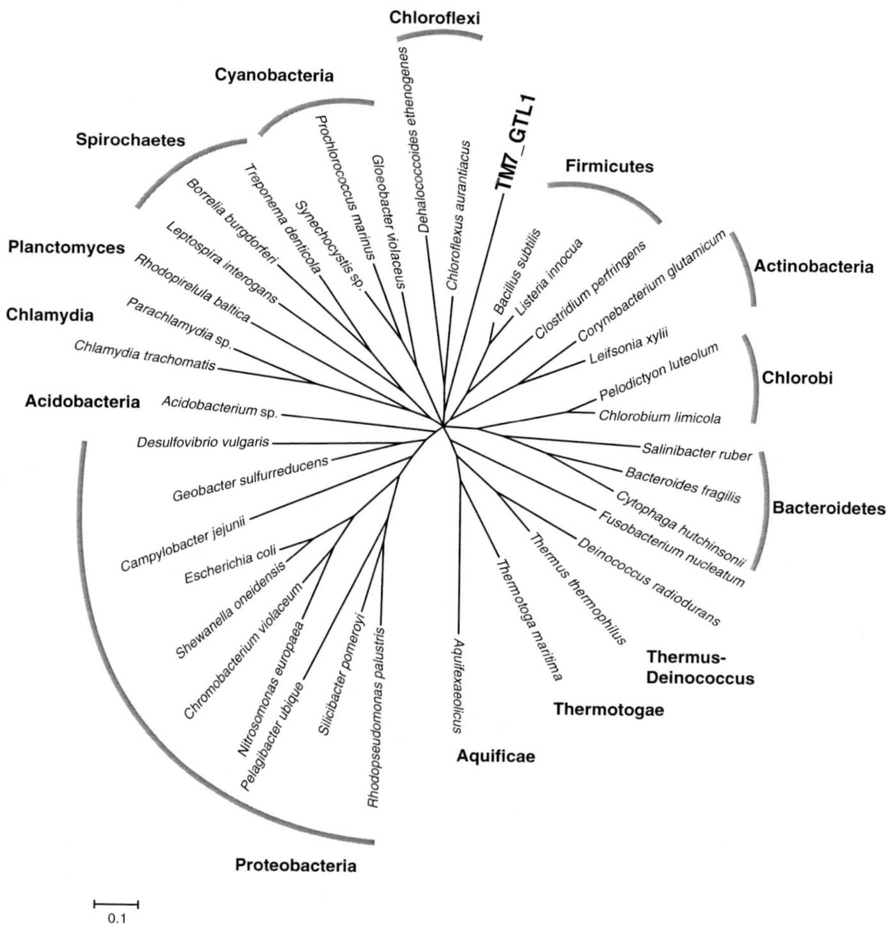

Fig. 2 Phylogenetic reconstruction of the relationships between bacterial phyla (divisions) based on concatenated ribosomal protein sequences. Numbers indicate bootstrap support, and branch-points with less than 70% bootstrap support were collapsed. Note the apparent relationship between TM7 and the Chloroflexi. Note also that the oral TM7 could not be added to this tree owing to an absence of many of the ribosomal proteins used in the analysis. Figure adapted from (Podar et al. 2007)

once since TM7 is not monophyletic with the Gram-positive phyla, the Actinobacteria, and Firmicutes, although further sequenced representatives of TM7 will be required to verify this inference. The Chloroflexi have unusual cell membranes (Hugenholtz and Stackebrandt 2004), and it will be interesting to further investigate the putative relationship to TM7 in this light when more genomic data become available. However, as it is often the case in reconstruction of the interphyla relationships within bacteria, the use of other phylogenetic markers leads to a relatively unresolved tree (Fig. 3).

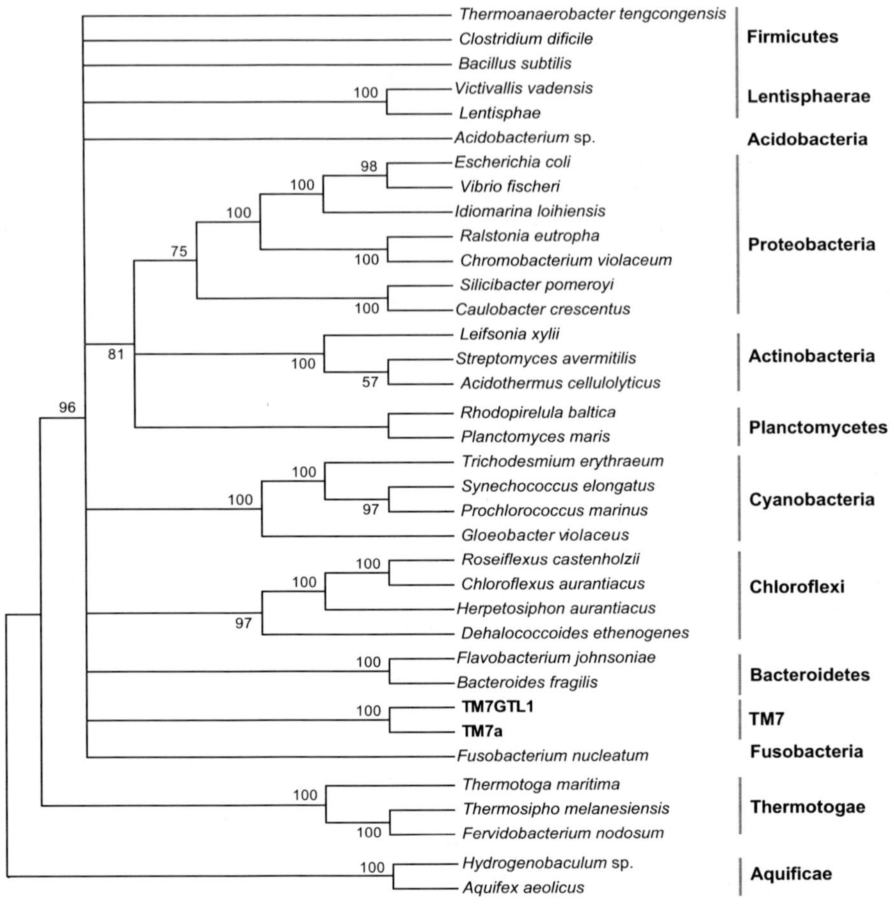

Fig. 3 Consensus dendrogram of the relationships between bacterial phyla (divisions) based on a concatenation of two conserved single copy marker genes: DNA gyrase (subunit A) and recA. Both genes were represented in the soil and oral TM7 datasets, allowing a comparative analysis of the two genomes beyond the 16S rRNA gene (Fig. 1). Branchpoints with less than 50% bootstrap support were collapsed. Note the monophyly of the TM7 genotypes but lower inter and intraphylum resolution as compared to the ribosomal proteins tree (Fig. 2), including loss of a specific relationship between TM7 and the Chloroflexi

Since we did not recover complete genomic sequences for either the soil or the human oral TM7 species, comprehensive metabolic reconstructions were not feasible. Nevertheless, there are certain aspects of the physiology and ecology of these organisms that can be inferred with reasonable confidence. Both TM7 genotypes contain genes involved in type IV pilus assembly, which has been implicated in twitching motility and biofilm formation, behavior that is important in colonizing either soil interstitial spaces or the tooth surface. A distinctive feature of the soil

TM7 appears to be an overrepresentation of restriction–modification systems, DNA repair, and antibiotic or heavy-metal transporters. These represent potentially important adaptations to an environment that is characterized by a very high diversity of microorganisms, competition, and predation as well as microbial chemical warfare. Unfortunately, the overlap of the two TM7 gene inventories was insufficient for inferring potential metabolic similarities and differences, owing presumably to the low and random genome coverage in both cases. However, as the remaining methodological hurdles are overcome in whole genome amplification and sequencing of single cells, a more complete picture of the metabolic repertoire of the TM7 phylum will emerge. More broadly speaking, it is clear that microbial ecology and evolution are entering a golden age and with the aid of powerful tools such as single cell genomics we will come to know the organisms with which we share the planet.

References

Abulencia CB et al. (2006) Environmental whole-genome amplification to access microbial populations in contaminated sediments. *Appl Environ Microbiol* 72:3291–3301

Amann R, Fuchs BM, Behrens S (2001) The identification of microorganisms by fluorescence in situ hybridisation. *Curr Opin Biotechnol* 12:231–236

Blanco L, Bernad A, Lazaro JM, Martin G, Garmendia C, Salas M (1989) Highly efficient DNA synthesis by the phage phi 29 DNA polymerase. *Symmetrical mode of DNA replication. J Biol Chem* 264:8935–8940

Brinig MM, Lepp PW, Ouverney CC, Armitage GC, Relman DA (2003) Prevalence of bacteria of division TM7 in human subgingival plaque and their association with disease. *Appl Environ Microbiol* 69:1687–1694

Cho JC, Vergin KL, Morris RM, Giovannoni SJ (2004) Lentisphaera araneosa gen. nov., sp. nov, a transparent exopolymer producing marine bacterium, and the description of a novel bacterial phylum, Lentisphaerae. *Environ Microbiol* 6:611–621

Ciccarelli FD, Doerks T, von Mering C, Creevey CJ, Snel B, Bork P (2006) Toward automatic reconstruction of a highly resolved tree of life. *Science* 311:1283–1287

Dean FB et al. (2002) Comprehensive human genome amplification using multiple displacement amplification. *Proc Natl Acad Sci USA* 99:5261–5266

DeSantis TZ et al. (2006) Greengenes, a chimera-checked 16S rRNA gene database and workbench compatible with ARB. *Appl Environ Microbiol* 72:5069–5072

Ferrari BC, Binnerup SJ, Gillings M (2005) Microcolony cultivation on a soil substrate membrane system selects for previously uncultured soil bacteria. *Appl Environ Microbiol* 71:8714–8720

Fu AY, Chou HP, Spence C, Arnold FH, Quake SR (2002) An integrated microfabricated cell sorter. *Anal Chem* 74:2451–2457

Garcia Martin H et al. (2006) Metagenomic analysis of two enhanced biological phosphorus removal (EBPR) sludge communities. *Nat Biotechnol* 24:1263–1269

Goldberg SM et sal. (2006) A Sanger/pyrosequencing hybrid approach for the generation of high-quality draft assemblies of marine microbial genomes. *Proc Natl Acad Sci USA* 103:11240–11245

Hawkins TL, Detter JC, Richardson PM (2002) Whole genome amplification--applications and advances. *Curr Opin Biotechnol* 13:65–67

Hong JW, Studer V, Hang G, Anderson WF, Quake SR (2004) A nanoliter-scale nucleic acid processor with parallel architecture. *Nat Biotechnol* 22:435–439

Hugenholtz P (2002) Exploring prokaryotic diversity in the genomic era. Genome Biol 3:REVIEWS0003

Hugenholtz P, Stackebrandt E (2004) Reclassification of Sphaerobacter thermophilus from the subclass Sphaerobacteridae in the phylum Actinobacteria to the class Thermomicrobia (emended description) in the phylum Chloroflexi (emended description). *Int J Syst Evol Microbiol* 54:2049–2051

Hugenholtz P, Goebel BM, Pace NR (1998) Impact of culture-independent studies on the emerging phylogenetic view of bacterial diversity. *J Bacteriol* 180:4765–4774

Hugenholtz P, Tyson GW, Webb RI, Wagner AM, Blackall LL (2001) Investigation of candidate division TM7, a recently recognized major lineage of the domain Bacteria with no known pure-culture representatives. *Appl Environ Microbiol* 67:411–419

Hutchison CA III, Venter JC (2006) Single-cell genomics. *Nat Biotechnol* 24:657–658

Konneke M, Bernhard AE, de la Torre JR, Walker CB, Waterbury JB, Stahl DA (2005) Isolation of an autotrophic ammonia-oxidizing marine archaeon. *Nature* 437:543–546

Kumar PS, Griffen AL, Barton JA, Paster BJ, Moeschberger ML, Leys EJ (2003) New bacterial species associated with chronic periodontitis. *J Dent Res* 82:338–344

Kvist T, Ahring BK, Lasken RS, Westermann P (2007) Specific single-cell isolation and genomic amplification of uncultured microorganisms. *Appl Microbiol Biotechnol* 74:926–935

Lasken RS, Stockwell TB (2007) Mechanism of chimera formation during the Multiple Displacement Amplification reaction. *BMC Biotechnol* 7:19

Liolios K, Tavernarakis N, Hugenholtz P, Kyrpides NC (2006) The Genomes On Line Database (GOLD) v.2: a monitor of genome projects worldwide. *Nucleic Acids Res* 34:D332–D334

Loy A, Maixner F, Wagner M, Horn M (2007) probeBase–an online resource for rRNA-targeted oligonucleotide probes: new features 2007. *Nucleic Acids Res* 35:D800–D804

Ludwig W et-al. (2004) ARB: a software environment for sequence data. *Nucleic Acids Res* 32:1363–1371

Marcy Y et al. (2007) Dissecting biological "dark matter" with single-cell genetic analysis of rare and uncultivated TM7 microbes from the human mouth. *Proc Natl Acad Sci USA* 104:11889–11894

Markowitz VM et-al. (2008) IMG/M: a data management and analysis system for metagenomes. *Nucleic Acids Res* 36:D534–D538

Mavromatis K et al. (2007) Use of simulated data sets to evaluate the fidelity of metagenomic processing methods. *Nat Methods* 4:495–500

McHardy AC, Martin HG, Tsirigos A, Hugenholtz P, Rigoutsos I (2007) Accurate phylogenetic classification of variable-length DNA fragments. *Nat Methods* 4:63–72

Ouverney CC, Armitage GC, Relman DA (2003) Single-cell enumeration of an uncultivated TM7 subgroup in the human subgingival crevice. *Appl Environ Microbiol* 69:6294–6298

Pinard R et al. (2006) Assessment of whole genome amplification-induced bias through high-throughput, massively parallel whole genome sequencing. *BMC Genomics* 7:216

Podar M et al. (2007) Targeted access to the genomes of low-abundance organisms in complex microbial communities. *Appl Environ Microbiol* 73:3205–3214

Preston CM, Wu KY, Molinski TF, DeLong EF (1996) A psychrophilic crenarchaeon inhabits a marine sponge: Cenarchaeum symbiosum gen. nov., sp. nov. *Proc Natl Acad Sci USA* 93:6241–6246

Rappe MS, Giovannoni SJ (2003) The uncultured microbial majority. *Annu Rev Microbiol* 57:369–394

Riesenfeld CS, Schloss PD, Handelsman J (2004) Metagenomics: genomic analysis of microbial communities. *Annu Rev Genet* 38:525–552

Ronaghi M (2001) Pyrosequencing sheds light on DNA sequencing. *Genome Res* 11:3–11

Speit G, Schutz P, Merk O (2000) Induction and repair of formaldehyde-induced DNA-protein crosslinks in repair-deficient human cell lines. *Mutagenesis* 15:85–90

Strous M et al. (2006) Deciphering the evolution and metabolism of an anammox bacterium from a community genome. *Nature* 440:790–794

Sundquist A, Ronaghi M, Tang H, Pevzner P, Batzoglou S (2007) Whole-genome sequencing and assembly with high-throughput, short-read technologies. PLoS ONE 2:e484

Thomsen TR, Kjellerup BV, Nielsen JL, Hugenholtz P, Nielsen PH (2002) In situ studies of the phylogeny and physiology of filamentous bacteria with attached growth. *Environ Microbiol* 4:383–391

Tringe SG, Rubin EM (2005) Metagenomics: DNA sequencing of environmental samples. *Nat Rev Genet* 6:805–814
Tyson GW et al. (2004) Community structure and metabolism through reconstruction of microbial genomes from the environment. *Nature* 428:37–43
Tyson GW, Lo I, Baker BJ, Allen EE, Hugenholtz P, Banfield JF (2005) Genome-directed isolation of the key nitrogen fixer Leptospirillum ferrodiazotrophum sp. nov. from an acidophilic microbial community. *Appl Environ Microbiol* 71:6319–6324
Woese CR (1987) Bacterial evolution. *Microbiol Rev* 51:221–271
Wolf YI, Rogozin IB, Grishin NV, Koonin EV (2002) Genome trees and the tree of life. *Trends Genet* 18:472–479
Xia Y, Kong Y, Nielsen PH (2007) In situ detection of protein-hydrolysing microorganisms in activated sludge. *FEMS Microbiol Ecol* 60:156–165
Xia Y, Kong Y, Thomsen TR, Nielsen PH (2008) Identification and ecophysiological characterization of epiphytic protein-hydrolyzing saprospiraceae (Candidatus Epiflobacter spp.) in activated sludge. *Appl Environ Microbiol* 74:2229–2238
Zengler K et al. (2005) High-throughput cultivation of microorganisms using microcapsules. *Methods Enzymol* 397:124–130
Zhang H et al. (2003) Gemmatimonas aurantiaca gen. nov., sp. nov., a gram-negative, aerobic, polyphosphate-accumulating micro-organism, the first cultured representative of the new bacterial phylum Gemmatimonadetes phyl. nov. *Int J Syst Evol Microbiol* 53:1155–1163
Zhang K et al. (2006) Sequencing genomes from single cells by polymerase cloning. *Nat Biotechnol* 24:680–686
Zoetendal EG, Plugge CM, Akkermans AD, de Vos WM (2003) Victivallis vadensis gen. nov., sp. nov., a sugar-fermenting anaerobe from human faeces. *Int J Syst Evol Microbiol* 53:211–215

Physiological and Ecological Adaptations of Slow-Growing, Heterotrophic Microbes and Consequences for Cultivation

Thomas M. Schmidt (✉) and Allan E. Konopka

Contents

Abstract There is a large discrepancy between the number of microbes that can be visualized in samples from most natural environments and the small number that grows readily in the laboratory. This anomaly hinders opportunities to advance our understanding of the vast metabolic and evolutionary diversity of microbes, and imposes severe limitations on our capacity to link patterns of ecological diversity with the functioning of microbial communities. This chapter focuses on slow-growing, heterotrophic microbes as a potential source of cultures to represent the remarkable phylogenetic diversity of the microbial world. Despite the obvious advantages conferred upon microbes that leave the most progeny per unit time, chronic limitation of nutrients in many environments selects for microbes that are able to survive and use resources efficiently, even if it involves a trade-off for faster growth when resources

T.M. Schmidt
Department of Microbiology and Molecular Genetics, Michigan State University, East Lansing, MI 48824, USA
e-mail: tschmidt@msu.edu

A.E. Konopka
Biological Sciences Division, Pacific Northwest National Laboratory, Richland, WA 99352, USA
e-mail: allan.konopka@pnl.gov WA 99352, USA

Microbiol Monogr, doi:10.1007/7171_2008_11

are abundant. Understanding the ecological strategies of slow-growing microbes and adjusting expectations for cultivation to match the physiological capabilities of these microbes offer an opportunity to narrow the tremendous gap between the microscopically visible microbes and those that are readily cultivated.

1 Introduction

The abundance and diversity of microbes on Earth is staggering: the biomass of bacteria and archaea rivals that in plants and animals (Whitman et al. 1998), and seemingly mundane environments, such as a gram of soil, are estimated to house at least thousands (Torsvik et al. 1998) and perhaps more than a million species of bacteria (Gans et al. 2005). Knowledge of the composition, organization, and fluctuations of indigenous microbial populations is scarce, even though the metabolism of microbes drives many ecosystem and global-level processes. One reason for the paucity of understanding of these microbes and microbial communities has been the general inability to cultivate the majority of microbes visible in microscopic examinations of soil – an irony referred to as “the great plate count anomaly” (Staley and Konopka 1985). Molecular surveys and metagenomic studies continue to document the tremendous genetic diversity of microbes and microbial communities, and these data form the basis for metabolic and ecological models of microbial life. To test and advance these models, microbiologists generally turn to laboratory cultures of microbes, where a multitude of chemical and physical parameters influencing the growth and survival of microbes can be manipulated.

The microbes that first caught the attention of biologists and continue to be the focus of the most intensive investigations are those that grow quickly, i.e., those that form visible colonies overnight on an agar surface. To form a visible colony after 12 h of incubation, microbes must have a doubling time of an hour or less. These fast-growing microbes are ideal for laboratory studies, where assessing a range of metabolic activities or screening for mutants or transformants can be accomplished following overnight incubation. For more slowly growing microbes, for instance one with a doubling time of 48 h, 5 weeks of exponential growth would be required to visualize a colony originating from a single cell (Leadbetter 2003). These longer doubling times are not uncommon: many aerobic, heterotrophic bacteria have doubling times such that weeks are required to visualize colonies (Fig. 1).

The simple strategy of extended incubation times has proven successful for cultivating representatives of previously uncultivated groups of bacteria including representative of the phyla Acidobacteria and Verrucomicrobia (Janssen et al. 2002; Stevenson et al. 2004; Eichorst et al. 2007). However, simply extending incubation times has not yielded the diversity of microbes that has been identified in molecular surveys. In order to devise better strategies for the cultivation of these bacteria from natural habitats, it seems valuable to consider the conditions under which they grow in nature and their physiological state prior to cultivation attempts. In this chapter, we focus on slow-growing, heterotrophic bacteria by providing some insights on

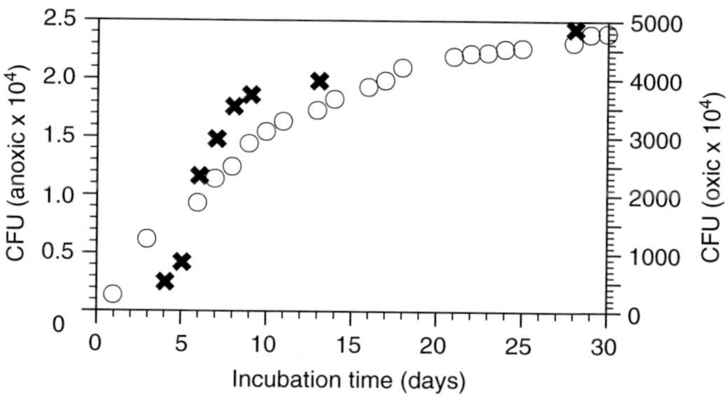

Fig. 1 Cumulative number of colonies following inoculation of an agar medium supplemented with low concentrations of complex carbon and nitrogen sources. Plates were incubated at room temperature under oxic(○) or anoxic(X) conditions. The plates were inoculated with a soil slurry from soils in a deciduous forest at the Kellogg Biological Station Long Term Ecological Research Site

their niches in nature and how physiological and ecological perspectives might help to inform successful cultivation strategies.

2 Resource- and Nutrient-Limited Growth

Microbial growth in nature is often characterized as "nutrient limited." This term embodies two distinct concepts: (1) the amount of biomass in the system is constrained by some nutrient in limited supply and (2) the concentration of this limiting nutrient has been reduced by bacterial assimilation to a level that limits growth rate to a "slow" value.

2.1 Nutrient Constraints on Biomass Level

Sprengel and Liebig were German agricultural scientists who promulgated and popularized the Law of the Minimum. This states that biomass yield is proportional to the mass of the essential nutrient that is present in the smallest supply relative to its requirement for biomass synthesis. This agronomic principle has been broadly applied in ecology; "bottom-up control" in ecosystems refers to cases where nutrient supply limits primary productivity.

In principle, any essential resource (energy source (electron donor), terminal electron acceptor, macronutrient (C, N, P, or S), or micronutrient (Fe and other metal ions)) could limit the amount of biomass in a system. However, as a practical matter, energy source limitations are considered most common for chemoheterotrophic and chemoautotrophic microbes (Cole et al. 1988; Kirchman 1990; Alden et al. 2001). For photoautotrophic microbes (cyanobacteria and eukaryotic algae), P is often limiting in freshwater and coastal marine aquatic habitats, whereas N is frequently limiting in the open ocean. Photosynthetic sulfur bacteria represent a special case (Parkin and Brock 1980), as their requirements for light, anaerobic conditions, and reduced S as electron donor constrain their habitats. Specific terminal electron acceptors (O_2, NO_3^-, SO_4^{-2}, and oxidized Fe and Mn) may become completely consumed and thus limit the abundance of specific physiological groups, but their hierarchy of usage (Fenchel et al. 1998) produces a succession of microbial types rather than a limit on biomass levels in a habitat.

Among the micronutrients, Fe deserves special consideration because the ratio between its cellular concentration and aqueous solubility (at neutral pH) is the largest of any nutrient (Brock 1966). Iron limitation of phytoplankton productivity in the Southern Ocean has been the subject of study for the past 10 years (Martin and Fitzwater 1988; Coale et al. 1996). Microbes have evolved specific mechanisms to mobilize Fe from the environment (Wandersman and Delepelaire 2004). Dinitrogen-fixing and metal-reducing bacteria may experience high demand for Fe, due to their high content of Fe-containing proteins that mediate these processes.

2.2 Nutrient Constraints on Growth Rate

If a microbial habitat is subject to bottom-up control and it is an open system with a continuous input of nutrients (due either to inflows from outside the system or recycling from within the system), then microbial growth will deplete an essential nutrient to the point where it not only controls the total amount of biomass, but also limits the growth rate (to a "slow" value). When growth rate of a pure culture is plotted as a function of substrate concentration (*S*), the general result is that specific growth rate (μ) is approximately proportional to *S* at low substrate concentrations, but reaches a plateau at higher substrate concentrations.

This type of saturation curve can be mathematically modeled by a variety of mathematical formulae (Jassby and Platt 1976; Koch 1997). What these formulae have in common are the nature of two parameters – one describes the effectiveness of growth at low concentrations and the other indicates the maximum possible growth rate (under the prevailing environmental conditions). The equation used most often in microbiology is that of Jacques Monod:

$$\mu = \mu_{\max} (S/(K_\mu + S)), \quad (1)$$

where μ_{max} is the maximum growth rate observed at saturating substrate concentrations, and $\boldsymbol{K}_\mu$ is the substrate concentration at which μ equals $0.5\mu_{max}$. This equation is analogous to the Michaelis–Menten equation used to describe the effect of substrate concentration upon the velocity of an enzymatic reaction.

Cultivation of bacteria in a chemostat illustrates the dynamic interplay between substrate flux, microbial growth rate, and biomass level in natural habitats, because both the chemostat and many natural habitats are open systems. Fig. 2 presents the results of a mathematical simulation of the start-up of a chemostat, a system in which there is continuous inflow of fresh nutrient medium into a culture vessel of

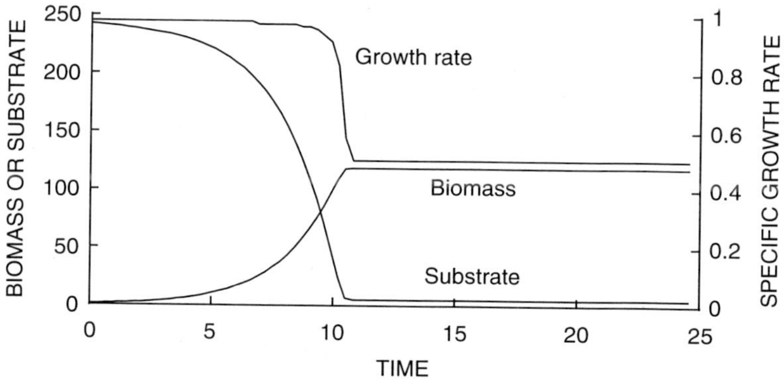

Fig. 2 Mathematical simulation of the start-up of a chemostat. The inoculated microbe had a μ_{max} of 1 h^{-1}, a $\boldsymbol{K}_m$ of 5 units of substrate. The chemostat was run with a dilution rate (***D***) of 0.5^{-1} (= 50% of μ_{max}), and the concentration of limiting substrate in the inflow reservoir was 250 units. The bacteria initially grow at an unrestricted rate near μ_{max} (and the population increases exponentially) until the substrate concentration is reduced. By definition, at steady state the growth rate balances the loss of bacteria because of dilution ($\mu = \boldsymbol{D}$). As ***D*** is set at one-half of μ_{max}, the required external substrate concentration is $\boldsymbol{K}_m$ (5 units)

constant volume (and continuous removal of culture at the same rate). Concentration of the limiting substrate is initially high, so the microbes grow at their μ_{max}. The dynamics appears similar to a batch culture growth curve in which biomass reaches a plateau when the cells reach stationary phase. However, there is an important difference. Cells are continuously washed out of the system in the effluent. Constancy in biomass means that the cell growth rate precisely balances the rate of loss by washout (the dilution rate). This growth rate is slow in comparison to the microbe's potential (μ_{max}). But how does the system become poised so that μ precisely equals dilution rate? Substrate concentration exerts this effect (as articulated in the Monod equation) and feedbacks between substrate concentration, growth rate, and microbial substrate consumption push the system back to the steady state if either steady-state biomass level or substrate concentration is perturbed.

Microbes are exquisitely efficient in maintaining high growth rates down to very low external substrate concentrations. In either natural habitats or chemostat cultures, microbes challenge the abilities of analytical chemists to measure the concentrations of limiting nutrients. The relatively small number of values reported for chemostat cultures (Button 1985; Button 1998) range from the low micromolar down to nanomolar concentrations. In situ concentrations of amino acids, sugars, and organic acids are reported to be in this range (Lovley and Klug 1982; Kirchman et al. 2001; Rosenstock and Simon 2001). Thus, even "dilute" microbiological media contain nutrient concentrations much higher than those experienced in most natural habitats.

The growth rate of bacteria in most natural habitats is generally "slow." Whereas doubling times of 1–3 h in chemically defined media are not unexpected for a variety of chemoheterotrophic bacteria in the laboratory, most estimates of doubling times for chemoheterotrophs in aquatic ecosystems cluster in the range of 18–50 h

(Staley and Konopka 1985; Smith and Prairie 2004). There are sparser data from soil habitats, but measurement of rates of macromolecular synthesis suggested that turnover times of bacteria were on the order of days rather than hours (Baath 1998). A cautionary note is that these are bulk-average measures in which a microbial activity (such as DNA or protein synthesis) is normalized to the total biomass content of the system. If there are significant proportions of inactive microbial biomass, growth rates for active microbes will be underestimated. For example, faster growth by particular bacterial clades was measured in estuaries (Yokokawa et al. 2004).

3 Physiological Adaptations to Nutrient Limitation

The high efficiency at which microbes can maintain high growth rates down to nanomolar to micromolar substrate concentrations leads one to explore the physiological adaptations they make under these conditions. The actual growth constraint is that low external substrate concentration limits the rate at which membrane-bound permeases can transport that substrate into the cell. Competition among different microbes in a habitat for a low, limiting concentration of nutrient represents a strong selective force for physiological adaptations that maximize nutrient flux into the cell.

The physiological state that arises under these transport-limited conditions may be relevant to the cultivatibility of organisms from nature (see Sect. 4). The physiological adaptations that are observed maximize transport and downregulate assimilatory reactions (biosynthesis of metabolic intermediates and macromolecules).

Slow-growing bacteria are much smaller in volume than those growing rapidly (in large part because of the reduced cellular content of ribosomes). This alone enhances nutrient transport (a property of the cell surface) in relation to nutrient assimilation (a property of enzymes in the cytoplasmic volume) because the ratio of surface area to volume increases as cell diameter decreases (Fig. 3). Bacterial cells in aquatic habitats often have cell diameters ranging from 0.5 to 1 mm (Button, Robertson et al. 1996), and it is over this size range that surface area to volume ratio shows its sharpest change.

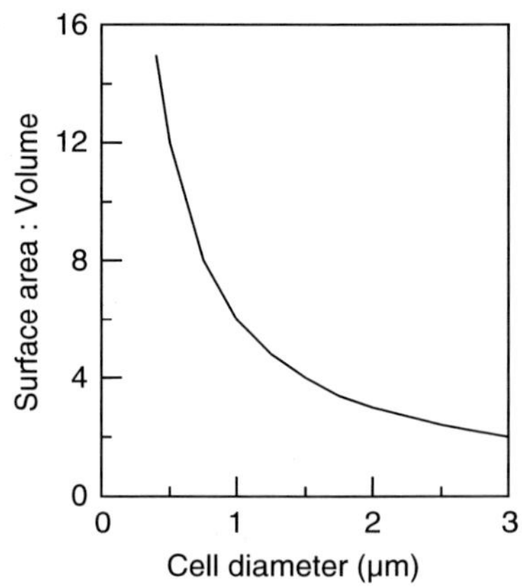

Fig. 3 Surface area to volume ratio for spherical cells of different diameters. Cells in aquatic habitats may have diameters as small as 0.4 μm

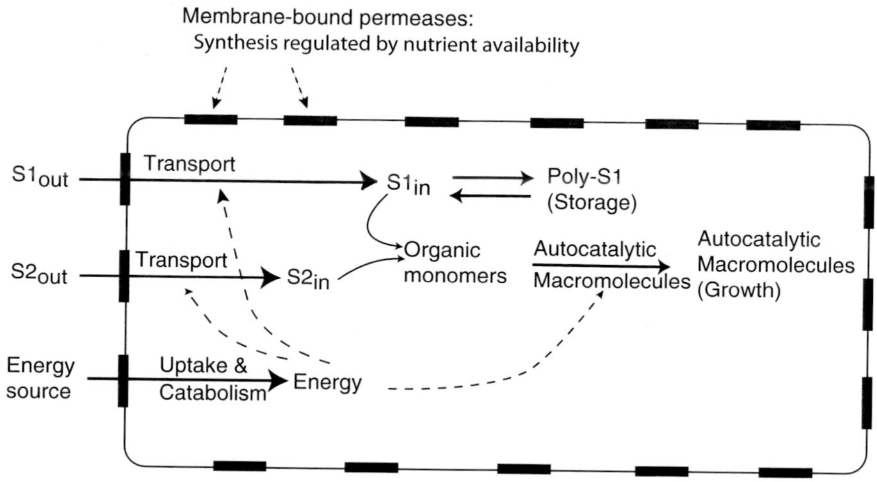

Fig. 4 Schematic representation of nutrient uptake and assimilation in bacterial cells. Membrane-bound permeases are responsible for transport of different nutrient resources (S1, S2, and energy source) from the exterior into the cell. Catabolic pathways oxidize the energy source and capture energy in biologically usable forms. Biosynthetic pathways convert nutrient resources into monomers needed for macromolecular synthesis. Of particular importance are autocatalytic macromolecules (proteins and nucleic acids) because they catalyze the synthesis of more macromolecules. If the rate of transport of a nutrient resource exceeds the rate of its assimilation, polymers (poly-S1) may accumulate for later utilization

In addition to the advantages that accrue with smaller cell size, slow-growing, nutrient-limited cells actively regulate gene expression and enzyme activities (Tempest et al. 1983; Koch 1997). These physiological adaptations (Fig. 4) include the following:

1. Maximize uptake of the rate-limiting element by derepressing the synthesis of membrane-bound permeases for that nutrient and (if present) inducing synthesis of new high-affinity permeases.
2. Derepress synthesis of permeases for alternative forms of the resource. For example, nitrate and amino acids represent alternative N sources to ammonia.
3. Downregulate the rate of cellular biosynthesis to match the inward flux of the limiting element. Lower concentrations of the monomers for macromolecular synthesis are maintained and there are lower cellular levels of "autocatalytic macromolecules", in particular ribosomes.
4. Storage polymers (glycogen, starch, or PHB for C and polyphosphate for P) are used to "average out" environmental heterogeneities in nutrient availability that may occur in either time or space. They provide a buffer capacity if a cell encounters a temporary excess supply of the limiting nutrient which it has the capacity to transport but not quickly assimilate for growth.

The most thorough theoretical analysis of how chemoheterotrophic microbes maximize the transport of limiting organic substrates has been carried out by Button

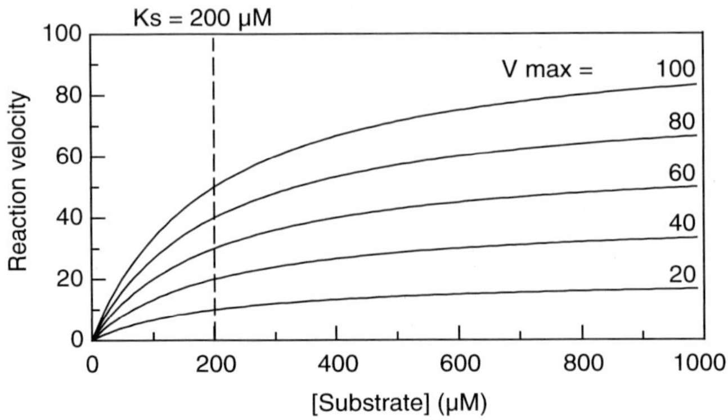

Fig. 5 The rate of nutrient resource transport by whole cells as a function of external concentration. The maximum rate of transport (V_{max}) increases as the number of transporters in the cell membrane is increased. Increasing the number of transporters also increases the flux rate of resource into the cell at low limiting concentrations

(Button 1994; Button et al. 1998). He considered the case of the ocean, in which labile dissolved organic carbon (DOC) limits bacterial populations. Although the sum total of DOC may be 1 mg l^{-1}, it consists of very low concentrations of many different individual organic molecules such as sugars, amino acids, and organic acids that arise from excretion or hydrolysis of macromolecules.

The rate of nutrient transport by whole cells is a nonlinear function of external substrate concentration (Fig. 5). The maximum rate (V_{max}) is related to the number of permease molecules studded on the cell membrane. Simple simulations illustrate that increasing permease content (seen as an increase in whole-cell V_{max}) increases the flux of nutrient into the cell at low external concentrations, even though the newly synthesized permeases have the same intrinsic binding capacity for substrate as the previous ones.

However, Button (Button 1994) reasoned that the flux of a single limiting organic substrate into the cell became maximal when a small fraction of the membrane was covered by its specific permease, on the basis of molecular collision frequency. Permeases with specificity for other organic substrates could be inserted into the remaining membrane area. Thus, an effective strategy for marine chemoheterotrophic bacteria when growth rate is limited by organic substrates is to express the capacity to transport multiple organic substrates; each substrate is present at very low concentrations. In fact, this is the only strategy that will allow them to accumulate sufficient energy and nutrients for growth at a doubling time of ca. 24 h when the concentrations of individual substrates are about 1 μg l^{-1} (Button et al. 2004).

Thus, when one attempts cultivation of microbes from nutrient-limited habitats, their physiological state should be taken into consideration. Cells are likely to have broad capacities for transport of a wide array of substrates, but very limited capacity to use energy and metabolic intermediates for biosynthesis.

4 Efficient Vs. Rapid Growth

After organic compounds are transported of into the cytoplasm of heterotrophic microbes, they enter into one of the central metabolic pathways and are either oxidized to carbon dioxide or incorporated into cell biomass. The capacity of respiring heterotrophic bacteria to convert organic compounds into biomass is known as ***Bacterial Growth Efficiency*** (BGE, and is defined as:

$$BGE = BP/(BP + BR) \quad (2)$$

where BP is bacterial production and BR is bacterial respiration. BGE is related to growth yield, but normalizes yield in such a manner that BGE ranges between 0 and 1. Maximal values of BGE are typically near 0.6, i.e., for every unit of carbon that is consumed, 0.6 units are incorporated into biomass and 0.4 units are respired to CO_2.

Variation in BGE is due largely to the relative amount of resource that is consumed to maintain cell viability. All bacteria spend some portion of their energy on processes that do not involve the production of new biomass, with the sum of these "non-growth" energy costs being referred to collectively as *maintenance energy*. When a bacterium is not doubling but still respiring, all of the energy consumed would be referred to as maintenance energy. A list of energy-consuming processes that contribute to maintenance energy is presented in Table 1. The amount of energy consumed as maintenance energy is also influenced by environmental conditions (Table 2), where for instance under nutrient-limited, slow-growth conditions, a greater percentage of resources is oxidized to support a cell's demand for maintenance energy.

In spatially heterogeneous environments typified by low resource availability, there is a fitness advantage conferred upon microbes that use resources efficiently, i.e., those with the highest BGE leave the most progeny per unit of substrate or

Table 1 Innate processes that influence a microbe's maintenance energy, which is defined as energy consumed for functions other than the production of new cell material

Process	Proposed mechanisms
Active transport	Direct consumption of ATP or ion motive force (IMF)
Motility	Consumption of IMF in flagellar motility or synthesis of excreted compounds in gliding motility
Translational machinery	Differences in the efficiency of protein synthesis
Pathway utilization	Differential consumption of ATP or reducing power based on selection of alternative metabolic pathways
Maintenance of ion gradients	Consumption of IMF or ATP to maintain internal pH or intracellular ion concentrations
Turnover of macromolecules	Half-life of proteins and other macromolecules dictates demand for new synthesis of these macromolecules
Stringent response	Synthesis of ppGpp
Yield from ATP synthase	Varying requirements for the number of protons required to generate ATP
Energy spilling	Oxidation of resources not coupled to energy generation or biosynthesis

Table 2 Environmental factors that may influence bacterial growth efficiency

Environmental Factor	Proposed mechanism
Free energy of substrates	Limits maximal ATP yield
Oxidation state of substrates	Cost of assimilatory reduction of C, S, and N for biosynthesis
Substrate concentration	Increase ATP demand for transport at low substrate concentrations; Energy spilling at excess concentrations
Availability of substrates	Demand for synthesis of compounds not available from environment
Deviation from a microbe's optimal growth conditions	Synthesis of macromolecules required for growth/survival in suboptimal conditions, e.g., detoxification of toxic compounds

simply survive the longest on limiting resources. And yet, cultivation conditions that do not provide for spatial separation of microbes, including most enrichment cultures, invariably select for the most rapidly growing microbes – the "weeds" of the microbial world. Under these conditions, there is no advantage conferred to organisms that use resources efficiently; rather the organism with the fastest generation time will become the most abundant organism in the culture.

The chemostat offers an alternative kind of enrichment culture, enriching for microbes able to harvest resources at low concentrations. It might be tempting to speculate that efficiency of resource utilization would be selected for under the nutrient-limited growth in chemostat culture; however, the organism that is most successful in a chemostat is simply the one that lowers the concentration of the limiting resource to a level where competitors are unable to acquire that compound. The equilibrium concentration to which an organism can lower the concentration of the limiting nutrient in a chemostat has been described in ecological terms to predict the outcome of competition experiment (Tilman 1981). That concentration is known as $\boldsymbol{R^*}$ and, when applied to a chemostat, is defined as follows:

$$R^* = sDK_{\mu}/(\mu_{max} - D), \quad (3)$$

where $\boldsymbol{D}$ is the dilution rate of the chemostat (also equal to the growth and loss rate since the population size is constant), $\boldsymbol{K}_m$ is the half-saturation constant for the limiting substrate, and μ_{max} is the maximum specific growth rate for a strain in that medium.

At $\boldsymbol{R^*}$, the concentration of the limiting nutrient supports a growth rate that is exactly sufficient to offset death by dilution, so the population size remains constant. Note that R^* depends on the Monod parameters, μ_{max} and $\boldsymbol{K}_s$, and on the experimentally controlled dilution rate, $\boldsymbol{D}$, but it does not depend on efficiency of resource utilization. In terms of competition, if species A draws down the limiting substrate to the concentration $\boldsymbol{R^*}_A$, which is below the corresponding equilibrium concentration of species B ($\boldsymbol{R^*}_B$), this means that populations of B cannot be sustained at this low substrate concentration and so species B cannot invade an established population of A. Increasing the concentration of the limiting resource in the reservoir results in a higher population density, but individual organisms in a perfectly mixed chemostat are exposed to the same physical and chemical environment, although at a higher cell density. Although chemostats select for populations that can harvest low concentrations of resources, it is μ_{max} and $\boldsymbol{K}_s$ that define the competitive outcome in a homogeneous environment, and not the efficiency of resource utilization.

When does efficiency provide a competitive advantage in nature? While characteristics that define the rate of growth (μ_{max} and $\boldsymbol{K}_m$ in (3)) have a major influence on the outcome of competition in homogeneous environments, many environments in nature – for example soils – provide discontinuous and temporally variable microenvironments where efficiency of resource utilization also comes into play. In these spatially segregated environments, increased efficiency of resource utilization can permit survival and growth under transient and low fluxes of resources. As the concentration of resources in the environment approaches zero, organisms must use internal stores of resources to survive, and again efficient use of resources is favored to survive periods without external inputs of nutrients. As will be discussed in the section on cultivation, there are several means to provide physical separation of microbes, and it is under these conditions that the slow-growing and efficient microbes have a competitive advantage.

A model based on a trade-off between the power and efficiency of the protein synthesizing machinery of bacteria was recently proposed (Dethlefsen and Schmidt 2007), and provides a mechanistic framework for understanding differences between rapid and efficient growth. Given that protein-synthesizing machinery is a major component of a cell's biomass and consumes more than half of the ATP equivalents generated by bacteria, the machinery of translation is undoubtedly key to understanding bacterial fitness in the environment. Slow-growing bacteria generally have few copies of the ribosomal RNA operon (Klappenbach et al. 2000), and the ribosomal RNA transcribed from those operons are incorporated into ribosomes that move slowly but are thought to make fewer errors (Dethlefsen and Schmidt 2007). Fast-growing bacteria are capable of manufacturing proteins more rapidly (per unit rRNA) and this appears to be a critical component of their life history characteristics (Stevenson and Schmidt 2004). Although there are numerous axes required to describe a microbe's niche, the fundamental importance of protein synthesis in microbes and its relationship to BGE provides a unifying model that may help identify where bacteria lie along a spectrum from efficient to rapid growth.

5 Slow Growth as a Strategy to Increase Cultivability

In ecological theory, organisms are often categorized as "***r***-selected" or "***K***-selected"; these terms arose from island biogeography theory (MacArthur and Wilson 1967). The letters connote the parameters of the logistic growth equation favored by plant and animal ecologists. The intrinsic rate of increase (***r***) is equivalent to the microbiologists' μ_{max}; if resources are not limiting, there is selection for species with the highest growth rate. ***K*** is the carrying capacity of the habitat; "***K***-selected" implies that resources are limiting and organisms that compete most effectively for limiting resources are selected. Although these terms are rarely used in microbial ecology, the alternative ecological strategies they imply are connoted by other terms. Copiotrophs or opportunists are thought to have a large μ_{max} and respond to patches of nutrients they encounter due to spatial or temporal heterogeneities. Oligotrophs (Poindexter 1981; Schut et al. 1997) can compete well at low substrate concentrations (and may not grow at high concentrations) but have a low μ_{max}.

Most classical cultivation strategies are likely to detect ***r***-strategists. Liquid enrichment cultures are inoculated at low density with a mixture of organisms from a natural sample. The batch system contains high nutrient concentrations relative to the habitat. Success is declared immediately when the liquid becomes turbid. However, the dominant organism is not necessarily the one that was present in highest density in the inoculum but rather the one with the highest μ_{max} under the elective condition. On solid medium inoculated with both fast- and slow-growing bacteria, it is the large and fast-growing colonies that are seen and transferred.

In the light of these considerations, the use of slow growth as a cultivation strategy requires (a) initial substrate concentrations that approximate those found in nature, (b) spatial separation of microbes so that fast-growing microbes do not monopolize available resources, and (c) patience. These characteristics are embodied in the work of Button and Hattori, along with their collaborators.

In order to cultivate the most abundant bacteria from marine habitats (which they reasoned were oligotrophs), Button and colleagues (Button et al. 1993) serially diluted a seawater sample in filter-sterilized seawater until near-extinction and the organisms were grown in the absence of external nutrient enrichment. Over the course of several weeks, cell numbers increased to the level found in unfiltered seawater (10^5–10^6 per ml). Via this "extinction culturing" approach, it was estimated that up to 60% of the bacteria were viable. The addition of 5 mg organic C per liter to enrichments drastically reduced initial culturability.

Hattori (Hattori et al. 1997) described the kinetics with which colonies arising from different bacterial species would appear on solid media. On this basis, soil bacteria were divided into several groups on the basis of their growth rate (Hashimoto and Hattori 1989). Different phylogenetic groups were prevalent in different phases of the colony-forming curve obtained over a 10-day incubation period (Mitsui et al. 1997).

These investigations provided important insights into the relationship of slow growth to cultivatability, but also bring to light some technical problems that must be resolved. Growth of marine bacteria in filtered seawater produces cultures that are not visibly turbid; thus other more sensitive methods must be used to detect growth. Button et al. (1993) used fluorescent staining of nucleic acids, and flow cytometry or fluorescence microscopy can be used to detect cell number increases. In the case of growth on nutritionally poor solid substrates, optical microscopy can be used as an aid to discern very small colonies (Hamaki et al. 2005).

Over the past decade, a number of other investigators built upon these insights related to slow growth in order to cultivate a higher proportion and broader range of microbes from nature.

Button's extinction culturing resulted in the isolation and analysis of two oligotrophic bacterioplankton (Button 1985; Vancanneyt et al. 2001; Button et al. 1998).

However, Connon and Giovannoni (2002) recognized that high-throughput methods would be advantageous to more rapidly isolate strains and then screen for previously uncultured but broadly distributed bacterial groups. By using microtiter plates as growth chambers and a cell array to quickly detect cell growth from individual wells by fluorescence microscopy, the investigators were able to obtain isolates of four undescribed or previously uncultured bacterioplankton. This included isolation (Rappe et al. 2002) of a representative of a ubiquitous clade of marine bacterioplankton (Morris et al. 2002), SAR11. Culturability using this system averaged about 9%, and 10–23 doublings occurred over a 3-week incubation period. This technical approach has also been used to isolate other novel bacterioplankton (Cho and Giovannoni 2004).

Janssen and his collaborators have used a solid medium format and investigated a broad variety of conditions in order to isolate a variety of soil bacteria that had previously been difficult to cultivate (Sait et al. 2002; Burns et al. 2004; Schoenborn et al. 2004; Davis et al. 2005). Joseph et al. (2003) stated that "for many of the 'unculturable' groups … no great innovations are required and that the growth of these microorganisms in pure culture with simple media is straightforward and reproducible." They did apply these strategies: (a) the use of gellan gum rather than agar (which contains several hundred milligrams of sugars and amino acids (Schut et al. 1993)) as a solidifying agent, (b) dilute complex media, soil extract, or plant polysaccharides as energy sources, and (c) incubation times up to 3 months. These investigators also counsel persistence – under the loosely selective conditions they use, analysis of more isolates may yield representatives of novel clades. For example, screening of 1,200 soil isolates with a specific oligonucleotide probe yielded 14 cultures of Verrucomicrobia (Sangwan et al. 2005).

Other innovative techniques have been devised to mimic microbes' natural environments and improve cultivatibility. Invariably, these techniques result in slow growth of the microbes, and the mass of available nutrients is small enough that only microcolonies are formed. Several investigators have used these approaches with sediment or soil microbes. Kaeberlein et al. (2002) embedded bacteria from marine sediment in agar within a diffusion chamber. The chamber was incubated on the surface of sediment, and microcolonies formed within a week. Most of these microbes would not subsequently grow on standard microbiological media, but could be passaged by transfer to another diffusion chamber or by coculture with other organisms. A soil substrate membrane system (SSMS) was devised to cultivate bacteria from soil (Ferrari et al. 2005). Bacteria were removed from soil particles and deposited on polycarbonate membranes by filtration. The polycarbonate filters serve as a physical support and are deployed on top of a soil slurry, which acts as a nutrient source. After 10 days of incubation, microcolonies of ca. 200 cells (representing 7.5 divisions, or a doubling time of >1 day) were observed. Phylogenetic analysis of 16S rRNA genes obtained by amplifying these genes from total DNA on the membranes found that sequences were generally more closely related to those found from analysis of environmental samples than from cultivated organisms. Unfortunately, only one of eight microcolonies was successfully passaged to another SSMS and none grew on a traditional laboratory medium. Keller and colleagues (Zengler et al. 2002) have used microencapsulation of individual microbes in an agar-based emulsion to physically separate microbes from one another, followed by incubation under a low nutrient flux and finally flow cytometry to detect microdroplets containing microcolonies. This approach combines all of the aspects discussed above for successful cultivation of oligotrophs and yielded numerous phylogenetically distinct microbes from seawater and soil samples.

The success of cultivation strategies that mimic natural habitats suggests that cultivating the majority of microbes from nature is possible. These approaches have already achieved 10-fold or greater increases in cultivatibility over traditional methods, and we can anticipate further innovations that improve their efficiency. A necessary advance is to determine how to efficiently propagate microcolonies and determine whether it is possible to obtain larger cell masses in culture. Hahn et al. (2004) used an acclimatization method (sequential addition of increasing amounts of organic substrates) to coax microbes that initially grew in filtered lake water to grow on traditional media.

5.1 The Abundant Vs. the Rare

All the current approaches entail diluting the heterogeneous microbial community until a single cell is inoculated into a liquid environment, or a small number of cells are distributed over a solid surface. As a result, these techniques are more likely to reveal the organisms present in the highest relative abundance, but not those that are rare. A gram of fertile soil has been estimated to contain between 10,000 (Torsvik et al. 1998) and 1 million different microbial genomes (Gans et al. 2005). Sogin et al. (2006) used a massively parallel tag sequencing strategy to interrogate deep oceanic water and reported that although there are a few bacterial populations at high relative abundance, they found thousands of highly diverse, low-abundance operational taxonomic units (OTUs) in these samples and termed these *the rare biosphere*.

The isolation of highly abundant bacterioplankton taxa (Schut et al. 1993; Rappe et al. 2002) has been a major success for slow cultivation techniques. However, a major challenge is to devise strategies that will also access microbes present at lower relative abundance. Some potential approaches are listed below.

5.1.1 Brute Force

If one has a specific phylogenetic target in mind, a nucleic acid probe can be used to identify rare types within a collection of cultivated strains (Gich et al. 2005). For example, Sangwan et al. (2005) isolated 14 verrucomicrobia from soil by testing 1,200 isolates with a specific probe. A polymerase chain reaction (PCR) based surveillance method (plate wash PCR; (Stevenson et al. 2004)) can be applied to identify isolation plates that contain microbes of interest. Secondary isolation procedures can then be directed at replicates of those plates that gave positive results with the surveillance method. These methodologies benefit greatly from high-throughput methods for cultivating individual strains (Connon and Giovannoni 2002; Bruns et al. 2003). Future developments in the sensitive and rapid detection of specific microbial groups are needed. In addition to the application of nucleic acid probes, chemical analysis of small amounts of biomass by spectroscopy (Naumann et al. 1991; Kalasinsky et al. 2007) or mass spectrometry (Vargha et al. 2006) and fluorescence-activated cell sorting (FACS; Park et al. 2005; Vives-Rego et al. 2000) could play important roles in high-throughput analyses to identify novel strains.

5.1.2 Use Environmental Heterogeneity

It is not clear from an ecological perspective why natural samples contain a very high richness of different microbes present at very low relative abundance. However, one factor in spatially structured habitats may be a consequence of environmental heterogeneity that causes physical isolation. That is, habitat species richness is apparently high because collection of a large sample integrates many different microhabitats which are dominated by a very small number of distinct microbes. This is most likely to occur in unsaturated soil habitats (Zhou et al. 2002), but may also be a factor in aquatic habitats (Seymour et al. 2004). Becker et al. (2006) used Denaturing Gradient Gel Electrophoresis (DGGE) analysis of 16S rRNA amplicons to show that 150 mg portions of soil contained a small number of dominant phylotypes and that the distribution of these phylotypes varied at a spatial scale of a centimeter or less.

Therefore, several different strategies to fractionate the sample could be used to increase the probability of cultivating the rare biosphere. The most direct would be to use small sample sizes in which local microbial communities might differ significantly in composition from the bulk average composition. In addition, soil samples have been fractionated on the basis of particle size, and differences in community composition have been found (Sessitsch et al. 2001). If the physical or chemical properties of different taxa differ, they could be physically separated on that basis, for example, by differential density centrifugation (Inoue et al. 2007) or FACS.

6 Why is There a Great Plate Count Anomaly?

Staley and Konopka (1985) pointed out what had long been known in environmental microbiology – that viable plate counts recovered a small fraction of bacteria in a natural sample that could be counted by direct microscopic count. Direct analysis of nucleic acid sequences recovered from natural samples has revealed an enormous diversity of microbial taxa that had not been detected by cultivation techniques. A conclusion drawn from these observations was that most microbes in nature are “uncultivatable.” Perhaps a more accurate description would be to say that our current knowledge of their physiology precludes devising an effective cultivation strategy. However, other genetic and physiological factors may contribute to the phenomenon.

The techniques described here that recognize the slow growth of bacteria – using very low initial substrate concentrations and incubating cultures weeks or months – have allowed easier cultivation of representatives of some (but by no means all) clades recognized by analysis of environmental 16S rRNA sequences. In some cases, only very dilute cultures in liquid or small microcolonies on solid surfaces are formed. This leads to the question whether these organisms are obligate oligotrophs (alternatively termed oligophiles (Watve et al. 2000)). Schut et al. (1997) reviewed the data 10 years ago and disputed that obligate oligotrophs existed. They suggested that microbes taken directly from severely substrate-limited natural environments would not grow directly at high nutrient conditions, but this condition often was transient and upon subculture the isolates would grow at elevated nutrient concentrations. If this occurs, it is difficult to determine whether the switch is physiological (see below) or a consequence of selection for mutations that permit growth at high substrate concentrations. Over the last 10 years, there have been more reports of successful cultivation strategies for dilute cultures or microcolonies in which a failure to produce macroscale cultures is reported. If these hold true, another facet of the "Great Plate Count Anomaly" is to adjust our concept of cultivation to include oligophiles that cannot be grown to high cell densities expected by microbiologists.

A third aspect of the anomaly concerns the physiological state of microbes. As discussed above, extreme nutrient limitation results in a physiological response in which cells have a high capacity for transport of diverse substrates and a reduced capacity to productively consume energy and metabolic intermediates via biosynthesis. Could depositing cells in this physiological state onto media with nutrients concentrations that saturate transport rates damage them even if they otherwise would be fully capable of growing on this medium? Koch (2001) listed a series of hypotheses why oligotrophs might be killed by challenge with high nutrient concentrations.

The idea that microbes in nature enter a physiological state that precludes their direct growth on high-nutrient media is analogous to the concept of "viable but nonculturable" (VBNC) for known cultivatable organisms. One idea that has been put forward is that VBNC is a consequence of a metabolic imbalance in which catabolism outpaces anabolism, with the result that toxic products of oxygen metabolism irreversibly damage cells following transfer from stressed conditions to nutrient-rich conditions (Bloomfield et al. 1998). For example, irreversible inactivation of proteins via carbonylation occurs in non-growing *Escherichia coli* under oxidative stress (Nystrom 2001). It will be difficult to conclusively prove that a broad range of microbes in a natural sample are recalcitrant to cultivation because of their physiological state. However, it would be of interest to conduct a general analysis of carbonylated proteins and other damaged macromolecules after a nutrient upshift applied to a natural population to determine whether oxidative damage is widespread.

If the physiological state of many microbes in nature is the barrier to direct cultivation, there is the promise that appropriate resuscitation conditions can make them "cultivatable" on the macroscale. Initial incubation at low nutrient fluxes under slow growth conditions could alter physiological state, and acclimatization methods (Hahn et al. 2004) may lead to subsequent macroscale cultivation at high nutrient densities. On the other hand, there are reports of failure to passage micro-colonies or low-density liquid cultures on media with higher nutrient concentrations, and it may be that some organisms are oligophiles.

7 Summary

Oligotrophic microbes are experts at surviving and reproducing under conditions of chronically limiting resources, and their successful cultivation requires approaches informed by their physiological status and ecological niche. In the pursuit of these ubiquitous organisms, microbiologists may need to adjust expectations of what constitutes successful laboratory cultivation: rather than monitoring turbidity in liquid cultures or expecting robust colony growth on solid supports, more sensitive methods for detecting growth may be required. Understanding the physiology and ecology of slow-growing microbes will undoubtedly continue to reward cultivation efforts and may provide much needed cultivars of some of the vast phylogenetic diversity of microbes.

Acknowledgement This work was supported in part by a grant from the National Science Foundation [IOS 0421900 to TMS].

References

Alden L, Demoling F et al. (2001) Rapid method of determining factors limiting bacterial growth in soil. *Appl Environ Microbiol* 67(4):1830–1838

Baath E (1998) Growth rates of bacterial communities in soils at varying pH: a comparison of the thymidine and leucine incorporation techniques. *Microb Ecol* 36(3):316–327

Becker JM, Parkin T et al. (2006) Bacterial activity, community structure, and centimeter-scale spatial heterogeneity in contaminated soil. *Microb Ecol* 51(2):220–231

Bloomfield S F, Stewart G et al.(1998) The viable but non-culturable phenomenon explained. *Microbiology* 144 (Part 1):1–3

Brock TD (1966) Principles of microbial ecology. Prentice Hall, Englewood Cliffs, NJ

Bruns A, Hoffelner H et al. (2003) A novel approach for high throughput cultivation assays and the isolation of planktonic bacteria. *FEMS Microbiol Ecol* 45(2):161–171

Burns DG, Camakaris HM et al. (2004) Combined use of cultivation-dependent and cultivation-independent methods indicates that members of most haloarchaeal groups in an Australian crystallizer pond are cultivable. *Appl Environ Microbiol* 70(9):5258–5265

Button DK (1985) Kinetics of nutrient-limited transport and microbial-growth. *Microbiol Rev* 49(3):270–297

Button DK (1994) The physical base of marine bacterial ecology. *Microb Ecol* 28(2):273–285

Button DK (1998) Nutrient uptake by microorganisms according to kinetic parameters from theory as related to cytoarchitecture. *Microbiol Mol Biol Rev* 62(3):636–645

Button DK, Schut F et al. (1993) Viability and isolation of marine-bacteria by dilution culture – theory, procedures, and initial results. *Appl Environ Microbiol* 59(3):881–891

Button DK, Robertson BR et al. (1996) Microflora of a subalpine lake: bacterial populations, size and DNA distributions, and their dependence on phosphate. *FEMS Microbiol Ecol* 21(2):87–101

Button DK, Robertson BR et al. (1998) A small, dilute-cytoplasm, high-affinity, novel bacterium isolated by extinction culture and having kinetic constants compatible with growth at ambient concentrations of dissolved nutrients in seawater. *Appl Environ Microbiol* 64(11):4467–4476

Button DK, Robertson B et al. (2004) Experimental and theoretical bases of specific affinity, a cytoarchitecture-based formulation of nutrient collection proposed to supercede the Michaels-Menten paradigm of microbial kinetics. *Appl Environ Microbiol* 70(9):5511–5521

Cho JC, Giovannoni SJ (2004) Cultivation and growth characteristics of a diverse group of oligotrophic marine Gammaproteobacteria. *Appl Environ Microbiol* 70(1):432–440

Coale KH, Johnson KS et al. (1996) A massive phytoplankton bloom induced by an ecosystem-scale iron fertilization experiment in the equatorial Pacific Ocean. *Nature* 383(6600):495–501

Cole JJ, Findlay S et al. (1988) Bacterial production in fresh and saltwater ecosystems – a cross-system overview. *Mar Ecol Prog Ser* 43(1–2):1–10

Connon SA, Giovannoni SJ (2002) High-throughput methods for culturing microorganisms in very-low-nutrient media yield diverse new marine isolates. *Appl Environ Microbiol* 68(8):3878–3885

Davis KER, Joseph SJ et al. (2005) Effects of growth medium, inoculum size, and incubation time on culturability and isolation of soil bacteria. *Appl Environ Microbiol* 71(2):826–834

Dethlefsen L, Schmidt TM (2007) Performance of the translational apparatus varies with the ecological strategies of bacteria. *J Bacteriol* 189(8):3237–3245

Eichorst SA, Breznak JA et al. (2007) Isolation and characterization of soil bacteria that define *Terriglobus* gen. *nov., in the Phylum Acidobacteria. Appl Environ Microbiol* 73(8):2708–2717

Fenchel T, King GM et al. (1998) Bacterial biogeochemistry: the ecophysiology of mineral cycling. Academic, San Diego, CA

Ferrari BC, Binnerup SJ et al. (2005) Microcolony cultivation on a soil substrate membrane system selects for previously uncultured soil bacteria. *Appl Environ Microbiol* 71(12):8714–8720

Gans J, Wolinsky M et al. (2005) Computational improvements reveal great bacterial diversity and high metal toxicity in soil. *Science* 309(5739):1387–1390

Gich F, Schubert K et al. (2005) Specific detection, isolation, and characterization of selected, previously uncultured members of the freshwater bacterioplankton community. *Appl Environ Microbiol* 71(10):5908–5919

Hahn MW, Stadler P et al. (2004) The filtration-acclimatization method for isolation of an important fraction of the not readily cultivable bacteria. *J Microbiol Methods* 57(3):379–390

Hamaki T, Suzuki M et al. (2005) Isolation of novel bacteria and actinomycetes using soil-extract agar medium. *J Biosci Bioeng* 99(5):485–492

Hashimoto T, Hattori T (1989) Grouping of soil bacteria by analysis of colony formation on agar plates. *Biol Fertil Soils* 7(3):198–201

Hattori T, Mitsui H et al. (1997) Advances in soil microbial ecology and the biodiversity. *Antonie Van Leeuwenhoek Int J Gen Mol Microbiol* 72(1):21–28

Inoue K, Nishimura M et al. (2007) Separation of marine bacteria according to buoyant density by use of the density-dependent cell sorting method. *Appl Environ Microbiol* 73(4):1049–1053

Janssen PH, Yates PS et al. (2002) Improved culturability of soil bacteria and isolation in pure culture of novel members of the divisions Acidobacteria, Actinobacteria, Proteobacteria, and Verrucomicrobia. *Appl Environ Microbiol* 68(5):2391–2396

Jassby AD, Platt T (1976) Mathematical formulation of relationship between photosynthesis and light for phytoplankton. *Limnol Oceanogr* 21(4):540–547

Joseph SJ, Hugenholtz P et al. (2003) Laboratory cultivation of widespread and previously uncultured soil bacteria. *Appl Environ Microbiol* 69(12):7210–7215

Kaeberlein T, Lewis K et al. (2002) Isolating "uncultivable" microorganisms in pure culture in a simulated natural environment. *Science* 296(5570):1127–1129

Kalasinsky KS, Hadfield T et al. (2007) Raman chemical imaging spectroscopy reagentless detection and identification of pathogens: signature development and evaluation. *Anal Chem* 79(7):2658–2673

Kirchman DL (1990) Limitation of bacterial-growth by dissolved organic-matter in the sub-arctic Pacific. *Mar Ecol Prog Ser* 62(1–2):47–54

Kirchman DL, Meon B et al. (2001) Glucose fluxes and concentrations of dissolved combined neutral sugars (polysaccharides) in the Ross Sea and Polar Front Zone, Antarctica. *Deep-Sea Res. II. Topical Stud Oceanogr* 48(19–20):4179–4197

Klappenbach JA, Dunbar JM et al. (2000) rRNA operon copy number reflects ecological strategies of bacteria. *Appl Environ Microbiol* 66(4):1328–1333

Koch AL (1997) Microbial physiology and ecology of slow growth. *Microbiol Mol Biol Rev* 61(3):305–318

Koch AL (2001) Oligotrophs versus copiotrophs. *BioEssays* 23:657–661

Leadbetter JR (2003) Cultivation of recalcitrant microbes: cells are alive, well and revealing their secrets in the 21st century laboratory. *Curr Opin Microbiol* 6(3):274–281

Lovley DR, Klug MJ (1982) Intermediary metabolism of organic-matter in the sediments of a utrophic lake. *Appl Environ Microbiol* 43(3):552–560

MacArthur RH, Wilson EO (1967) The theory of island biogeography. Princeton University Press, NJ

Martin JH, Fitzwater SE (1988) Iron-deficiency limits phytoplankton growth in the northeast Pacific subarctic. *Nature* 331(6154):341–343

Mitsui H, Gorlach K et al. (1997) Incubation time and media requirements of culturable bacteria from different phylogenetic groups. *J Microbiol Methods* 30(2):103–110

Morris RM, Rappe MS et al. (2002) SAR11 clade dominates ocean surface bacterioplankton communities. *Nature* 420(6917):806–810

Naumann D, Helm D et al. (1991) Microbiological characterizations by FT-IR spectroscopy. *Nature* 351(6321):81–82

Nystrom T (2001) Not quite dead enough: on bacterial life, culturability, senescence, and death. *Arch Microbiol* 176(3):159–164

Park HS, Schumacher R et al. (2005) New method to characterize microbial diversity using flow cytometry. *J Ind Microbiol Biotechnol* 32(3):94–102

Parkin TB, Brock TD (1980) Photosynthetic bacterial production in lakes – the effects of light-intensity. *Limnol Oceanogr* 25(4):711–718

Poindexter JS (1981) Oligotrophy – fast and famine existence. *Adv Microb Ecol* 5:63–89

Rappe MS, Connon SA et al. (2002) Cultivation of the ubiquitous SAR11 marine bacterioplankton clade. *Nature* 418(6898):630–633

Rosenstock B, Simon M (2001) Sources and sinks of dissolved free amino acids and protein in a large and deep mesotrophic lake. *Limnol Oceanogr* 46(3):644–654

Sait M, Hugenholtz P et al. (2002) Cultivation of globally distributed soil bacteria from phylogenetic lineages previously only detected in cultivation-independent surveys. *Environ Microbiol* 4(11):654–666

Sangwan P, Kovac S et al. (2005) Detection and cultivation of soil verrucomicrobia. *Appl Environ Microbiol* 71(12):8402–8410

Schoenborn L, Yates PS et al. (2004) Liquid serial dilution is inferior to solid media for isolation of cultures representative of the phylum-level diversity of soil bacteria. *Appl Environ Microbiol* 70(7):4363–4366

Schut F, Devries EJ et al. (1993) Isolation of typical marine-bacteria by dilution culture – growth, maintenance, and characteristics of isolates under laboratory conditions. *Appl Environ Microbiol* 59(7):2150–2160

Schut F, Prins RA et al. (1997) Oligotrophy and pelagic marine bacteria: facts and fiction. *Aquat Microb Ecol* 12(2):177–202

Sessitsch A, Weilharter A et al. (2001) Microbial population structures in soil particle size fractions of a long-term fertilizer field experiment. *Appl Environ Microbiol* 67(9):4215–4224

Seymour JR, Mitchell JG et al. (2004) Microscale heterogeneity in the activity of coastal bacterioplankton communities. *Aquat Microb Ecol* 35(1):1–16

Smith EM, Prairie YT (2004) Bacterial metabolism and growth efficiency in lakes: the importance of phosphorus availability. *Limnol Oceanogr* 49(1):137–147

Sogin ML, Morrison HG et al. (2006) Microbial diversity in the deep sea and the underexplored "rare biosphere". *Proc Natl Acad Sci USA* 103(32):12115–12120

Staley JT, Konopka A (1985) Measurement of insitu activities of nonphotosynthetic microorganisms in aquatic and terrestrial habitats. *Annual Rev Microbiol* 39:321–346

Stevenson BS, Schmidt TM (2004) Life history implications of rRNA gene copy number in Escherichia coli. *Appl Environ Microbiol* 70(11):6670–6677

Stevenson BS, Eichorst SA et al. (2004) New strategies for cultivation and detection of previously uncultured microbes. *Appl Environ Microbiol* 70(8):4748–4755

Tempest DW, Neijssel OM et al. (1983) Properties and performance of microorganisms in laboratory culture: their relevance to growth in natural ecosystems. *Symp Soc Gen Microbiol* 34:119–152

Tilman D (1981) Tests of resource competition theory using four species of Lake Michigan algae. *Ecology* 62(3):802–815

Torsvik V, Daae FL et al. (1998) Novel techniques for analysing microbial diversity in natural and perturbed environments. *J Biotechnol* 64(1):53–62

Vancanneyt M, Schut F et al. (2001) Sphingomonas alaskensis sp nov., a dominant bacterium from a marine oligotrophic environment. *Int J Syst Evol Microbiol* 51:73–80

Vargha M, Takats Z et al. (2006) Optimization of MALDI-TOF MS for strain level differentiation of Arthrobacter isolates. *J Microbiol Methods* 66(3):399–409

Vives-Rego J, Lebaron P et al. (2000) Current and future applications of flow cytometry in aquatic microbiology. *FEMS Microbiol Rev* 24(4):429–448

Wandersman C, Delepelaire P (2004) Bacterial iron sources: from siderophores to hemophores. *Annu Rev Microbiol* 58:611–647

Watve M, Shejval V et al. (2000) The 'K' selected oligophilic bacteria: a key to uncultured diversity? *Curr Sci* 78(12):1535–1542

Whitman WB, Coleman DC et al. (1998) Prokaryotes: the unseen majority. *PNAS* 95(12):6578–6583

Yokokawa T, Nagata T et al. (2004) Growth rate of the major phylogenetic bacterial groups in the Delaware estuary. *Limnol Oceanogr* 49(5):1620–1629

Zengler K, Toledo G et al. (2002) Cultivating the uncultured. *PNAS* 99(24):15681–15686

Zhou JZ, Xia BC et al. (2002) Spatial and resource factors influencing high microbial diversity in soil. *Appl Environ Microbiol* 68(1):326–334

Viable but Not Cultivable Bacteria

Rita R. Colwell

Contents

Abstract A well-studied, long-term survival mechanism employed by Gram-positive bacteria is formation of endospores. For Gram-negative bacteria, the assumption has been that a survival state does not exist. However, a dormancy state has been described for Gram-negative bacteria and designated as the viable but nonculturable (VBNC) strategy of nonspore-forming cells. A variety of environmental factors are involved in induction of the viable but nonculturable state and *Vibrio cholerae* provides a useful paradigm for the VBNC phenomenon. It is now accepted that plate counts cannot be relied upon to enumerate or detect VBNC cells. Therefore, direct methods employing fluorescent staining, molecular genetic probes, and other molecular methods have proven both useful and reliable in detecting and enumerating both culturable and nonculturable cells. A predictive model for cholera has been developed, based on ground truth data gathered using these molecular methods and combining them with data obtained by remote sensing, employing satellites. It is clear that microbiology in the twenty-first century has been enhanced by these new tools and paradigms.

R.R. Colwell
University of Maryland, Institute for Advanced Computer Studies,
College Park, MD, 20742-0001, USA
e-mail: rcolwell@umiacs.umd.edu

Microbiol Monogr, doi:10.1007/7171_2009_1

1 Introduction

In response to environmental signals, many organisms enter a state of suspended animation, an extreme form of quiescence in which microscopically visible movement no longer continues. Nystul et al. (2003) have described this phenomenon, induced by anoxia in *Caenorhabditis elegans* embryos, and identified a gene, *san-1*, required for suspended animation, which functions as a spindle checkpoint component. Diapause, hibernation, and estivation are essentially programs that allow organisms to survive harsh seasonal conditions. In bacteria, a well-studied, long-term survival mechanism is formation of endospores, limited in general to Gram-positive bacteria. The assumption has been that nonspore-forming bacteria do not possess a survival mechanism to withstand harsh environments for long periods of time. The dormant-like state described by Xu et al. (1982) as viable but nonculturable (VBNC or VNC) represents a strategy of the nonspore-forming cells to accomplish that objective.

Starvation is a common stress inducing the VBNC state, either independently or in combination with other factors such as ultraviolet light, intense visible light, peracetic acid treatment, changes in osmolarity or aeration, toxic forms of copper, silver, as well as biological factors (Colwell and Grimes 2000). *Vibrio cholerae* has served as a useful example of VBNC, i.e., the noncultivable condition, since it is both an environmental bacterium and a pathogen, causing cholera, a pandemic disease in humans.

2 Viable but Nonculturable State in *Vibrio cholerae* and Other Species

Cholera is a devastating disease that takes thousands of lives each year worldwide and is endemic in many African, Asian, and Latin American countries, but predominantly so in the Indian subcontinent. In Bangladesh, there are two distinct seasonal peaks of cholera, one in spring and another in late fall (Sack et al. 2003). Cholera bacteria rarely can be isolated or even detected during interepidemic periods. Direct detection fluorescent staining, optimized using monoclonal antibody, was employed in an extensive study conducted in Bangladesh over a 3-year period, finding that >63% of plankton samples were positive for *V. cholerae* by direct detection using monoclonal antibody staining, while none were positive by culture (Huq et al. 1990). It was the first report demonstrating the presence of *V. cholerae* year round in Bangladesh water sources, with increased numbers of cells detected in the early spring and late fall, coinciding with the two peaks of cholera in that country. In earlier experiments, it had been demonstrated that when cells of *V. cholerae* attach to plankton, they often became noncultivable (Colwell and Huq 1994). A significant observation was that the number of noncultivable cells detected by molecular methods tended to increase, peaking just before a clinical case of cholera was reported, after which culturable *V. cholerae* could be isolated from the water (Huq and Colwell, unpublished data).

Thus, the concept of the viable but not cultivable state into which many bacterial species enter when confronted with adverse environmental conditions, was introduced 30 years ago, with *V. cholerae* serving as the prototype. The phenomenon is now known to be characteristic of many Gram-negative and also some nonsporulating Gram-positive bacterial species. Triggering factors include nutrition deprivation, exposure to sublethal concentrations of chemicals, antibiotics, or other antimicrobial agents, as well as environmental factors, notably temperature (Colwell 1996; Huq et al. 2005; Lobitz et al. 2000). When bacteria are not cultivable, they maintain a limited endogenous respiration, as well as their virulence and pathogenicity. Such cells commonly undergo significant reduction in size, become coccoid in shape, and are small enough to pass through a 0.45-μm pore size filter and even, in some cases, through a 0.22-μm filter (Chaiyanan 2007). Since the first publication of studies on the viable but not cultivable *V. cholerae*, a significant body of research has been done on a variety of genera, e.g., *Listeria, Helicobacter*, *Edwardsiella*, etc. (Colwell and Grimes 2000). Recently, Liu et al. (2008) have described a sensitive method for specific detection of viable *Escherichia coli* O157:H7 cells, including VBNC cells in water samples.

The method involves capture of bacterial cells on a low-protein-binding membrane and direct extraction and purification of RNA, followed by reverse transcription-PCR and electronic microarray detection of the *rfb E* and *fli C* genes of *E. coli* O157:H7. As few as 3–4 CFU l^{-1} in tap water, 7 CFU l^{-1} in river water, and 50 VBNC cells in 1 l of river water could be detected.

Depending on the efficiency and/or selectivity of the culture medium employed, interpretation of viability in bacterial cells varies significantly. If an organism did not grow on a given culture medium, it was considered not to be viable, that is, dead, when, in fact, the cells were viable, but simply unable to grow on the medium provided, usually for any one of a variety of reasons, such as depleted nutrient supply, lack of a specific nutrient required for growth, adverse physical or chemical conditions, or simply because dormancy was a part of the life cycle of the cell.

Life and death in bacteria has been debated for decades. Valentine and Bradfield (1954) proposed the term, viable, to describe cells capable of multiplying and forming colonies. However, the term, live, was used for respiring cells unable to divide under specified conditions. A decade later, Postgate and Hunter (1962) referred to cells as dead, when they did not divide and, with caution, termed such cells to be alive if they retained the osmotic barrier of the cell wall. The term dormancy has been employed in the literature by microbiologists, usually to describe spore and cyst-forming bacteria (Postgate 1969). The term moribund was suggested to describe a transient state of cells, i.e., a state between viability and death, usually applied to populations of cells exposed to starvation (Postgate 1967). Investigators (Kurath and Morita 1983) have also redefined the postulations of Valentine and Bradford (1954), proposing cells not capable of forming colonies to be nonviable. Spores and cysts produced by species of soil bacteria as survival mechanisms have been referred to as “resting” forms. Somnicell is a useful descriptive for bacterial cells entering into a state of existence, whereby culturing is not successful, but the cells continue to respire.

Investigators in the early 1980s compared methods for direct detection of bacterial cells, using *V. cholerae* and *Salmonella enteritidis* as test species. Roszak et al. (1984)

used direct detection methods to study cells failing to grow, or form colonies in or on bacteriological media, proposing the descriptor, viable but noncultivable (VBNC). To study these organisms, it was necessary, even though difficult, to determine the presence of these bacteria in samples collected from the natural environment. Polyclonal antibodies were employed initially to detect the cells, using an indirect fluorescent antibody staining method to determine both the presence and number in environmental samples (Xu et al. 1984). In the mid-1980s, the method was later optimized by employing monoclonal antibodies (Brayton and Colwell 1987). More recently, the gene expression profile of VBNC *V. cholerae*, induced by cold stress, has been studied (Asakura et al. 2006).

These investigators induced the VBNC state in *V. cholerae* O1 El Tor strain P6973 by incubation in artificial seawater at 4°C. The cells, after incubation for 70 days, retained membrane integrity and were pathogenic, colonizing the gut of iron–dextron-treated mice, even though unable to form colonies on laboratory media. The global transcription pattern of the VBNC cells was compared with that of stationary-phase cells. A total of 100 genes were induced more than fivefold in the VBNC state and the modulated genes were mostly those responsible for cellular processes. The changes in expression were verified by real-time RT-PCR analysis. The VC0230 [iron (111) ABC transporter], VC1212 (pol B), VC2132 (fli G) and VC2187 (fla C) m-RNAs were increased in the nonculturable state.

V. cholerae grows well on a thiosulfate–citrate–bile salts (TCBS) medium, giving rise to bright yellow colonies. The hypothesis that vibrios are autochthonous inhabitants of the marine and estuarine environment was initially rejected by many investigators in the early 1980s because, historically, *V. cholerae,* the causative agent of cholera, was considered to be strictly a human pathogen, transmitted only person-to-person (Colwell et al. 1981). Most importantly, the lack of proof of the existence of *V. cholerae* in the environment between epidemics of cholera was the main reason for such opposition. Today, the evidence is indisputable that *V. cholerae* and many other pathogens, e.g., *Helicobacter pylori* (Azevedo et al. 2007), can be present in a noncultivable state in the aquatic environment between epidemics, escaping detection unless appropriate technology is employed for detection, generally molecular genetic methods.

Robert Koch is credited with describing *V. cholerae*, and he also speculated that it multiplied in river water (Koch 1884). Unfortunately, attention was not given to further study of the ecology of *V. cholerae* until more than 100 years after the critical observations of John Snow in 1854 and Robert Koch in 1884. In the early 1970s, there was a renewed interest by microbial ecologists in the ecology of vibrios (Sochard et al. 1979). Reports of cholera in Australia in 1977 and in Texas in the United States in 1978, were believed, at first, to be imported cases. However, cholera subsequently was confirmed as indigenous in origin in both instances and generated a renewed interest in the ecology of microbial pathogens in the environment (Blake 1994).

Toxigenic forms of *V. cholerae* O1 were detected and isolated from samples of natural water collected from Louisiana bayous and from the Chesapeake Bay in 1980 (Colwell et al. 1981) and during this time, VBNC cholera vibrios were discovered to be a naturally occurring component of the aquatic environment, with wider acceptance of the uncultivable state of *V. cholerae* and related Gram-negative bacteria (Huq and Colwell 2000).

To understand the physiology and ecology of viable but noncultivable cells, cultivable cells of *V. cholerae* were successfully induced into the VBNC state in laboratory microcosms (Xu et al. 1984) and injected into rabbit ilial loops to determine persistence of virulence in VBNC cells (Colwell et al. 1985). A study designed to test whether VBNC cells of *V. cholerae* remained pathogenic and employing human volunteers was carried out (Colwell et al. 1996). From results of these studies, it was concluded that viable but noncultivable cells are capable of producing disease in animals and humans (Colwell et al. 1996). Retention of virulence in VBNC *Edwardsiella tarda* has been demonstrated by Du et al. (2007). Retention of plasmids also has been demonstrated (Byrd and Colwell 1990) and retention of enteropathogenicity by VBNC *E. coli* exposed to seawater and sunlight (Pommepuy et al. 1996). The latter is a significant observation in view of the report by Boyle et al. (2008) on batch solar disinfection (SODIS) having bactericidal properties. Whether SODIS induces the VBNC state has not been determined. In view of the retention of pathogenicity by many species of bacteria that have been found to have the capacity to enter a VBNC state, the public health application of SODIS may be limited.

A recently described emerging pathogen, *Arcobacter butzler*, causing diarrhea has been shown to be inducible to the VBNC state (Fera et al. 2008). Besides incubation at low temperatures, chemical treatment of bacteria at sublethal concentrations of antibacterial agents was found to induce conversion to the noncultivable state, but also to alter the phenotype (Chowdhury et al. 1997; McFeters and LeChevallier 2000).

Studies by Cappelier et al. (2007) have shown that VBNC *Listeria monocytogenes* require the presence of an embryo to be recovered in egg yolk and regain virulence after recovery, a very important finding since this species is a human pathogen.

The morphology and size of bacteria in the viable but noncultivable state are characteristic, i.e., cells are reduced in size, ranging from 0.01 to 0.2 μm in diameter, and coccoid (Folk 1996; Novitsky and Morita 1976). Chaiyanan et al. (2007) carried out microcosm experiments and electron microscopy, showing significant variation in shape and size of *V. cholerae* cells in the viable but not cultivable state (Chaiyanan 2007). Similar studies have been done with *H. pylori* (Azevedo et al. 2007).

3 Methods of Enumeration of Viable but Nonculturable Cells

It is now accepted that plate counts cannot be relied upon to enumerate or detect viable but noncultivable organisms and, therefore, the total bacterial population in a given sample. Alternative methods, based on respiration and metabolic activity measurement, direct and indirect enumeration employing immunological tests, and most importantly, molecular methods are required (Huq et al. 2000). A simple method, acridine orange direct count (AODC), has proven useful to obtain total cell counts (Hobbie et al. 1977). AODC, combined with direct viable counting, was developed in the late 1970s by Kogure et al. (1979) to enumerate viable but noncultivable cells, using a yeast extract enrichment step in the presence of nalidixic acid. During enrichment, active cells are identified by elongation and metabolism without replication or cell division. Direct fluorescent antibody staining, combined with the direct viable count method is now widely used in microbial ecology (Chowdhury et al. 1995).

Molecular methods to investigate viable but noncultivable cells include both hybridization with DNA or RNA probes and DNA amplification employing PCR (Knight 2000). Noncultivable *V. fischeri* has been found to be in abundance in sea water, producing symbiotic infections (Lee and Ruby 1995). PCR when used with appropriate gene probes allows detection of specific genes, and notably is useful in determining presence of toxigenic strains of bacteria. Confocal laser microscopy, in combination with fluorescent-based hybridization assays, provides a sensitive method for detecting and identifying viable but noncultivable microorganisms. Urografin density gradient centrifugation has been proposed to separate noncultivable cells from populations of *E. coli* (Arana et al. 2008) as well as flow cytometry (Vital et al. 2007). A limitation of nucleic-acid-based procedures for analysis of natural communities is that unbiased isolation of more or less pure and undegraded nucleic acid is critical.

VBNC *V. cholerae* O1 in biofilms have been suggested as an important factor in cholera transmission (Alam et al. 2007). Viable but noncultivable bacteria have been detected in human urine using the BacLight LIVE/DEAD (Molecular Probes, Inc., Eugene, Oregon) technique, suggesting that they may be the cause of chronic urinary tract infections. In light of the recent interest in the dormant persistent state of bacterial pathogens, following antibiotic treatment, and as related to latent bacterial infections and biofilm multidrug tolerance, the VBNC state appears to offer useful insight for these public health problems (Lewis 2007).

4 Modeling the Occurrence of Cholera

A predictive model for cholera has been developed, employing environmental, chemical, and biological data, both ground truth and satellite sensor data, i.e., remote sensing data obtained via satellite. Among parameters investigated in studies conducted in different geographical locations, temperature appears to be the most significant of environmental factors for predicting occurrence of cholera. Data obtained via satellite remote sensing in the Bay of Bengal showed direct correlation of cases of cholera in Matlab, Bangladesh with elevated sea surface temperature (Lobitz et al. 2000; Constanin de Magny et al. 2008). In one such completed study, a threshold of temperature at 25°C significantly increased the probability of occurrence of cholera in Bangladesh (Huq et al. 2005). That is, each 5° rise in water temperature increased the risk of cholera 3.3-fold. The model, based on the observed number of cholera cases versus number of cases predicted by the Poisson Regression model and upper 95% prediction limit, showed a 6-week lag between temperature rise and onset of cholera cases (see Fig. 1).

In conclusion, knowledge gained from studies of the ecology of VBNC *V. cholerae* and other bacterial species can provide better understanding of bacteria in their natural environment.

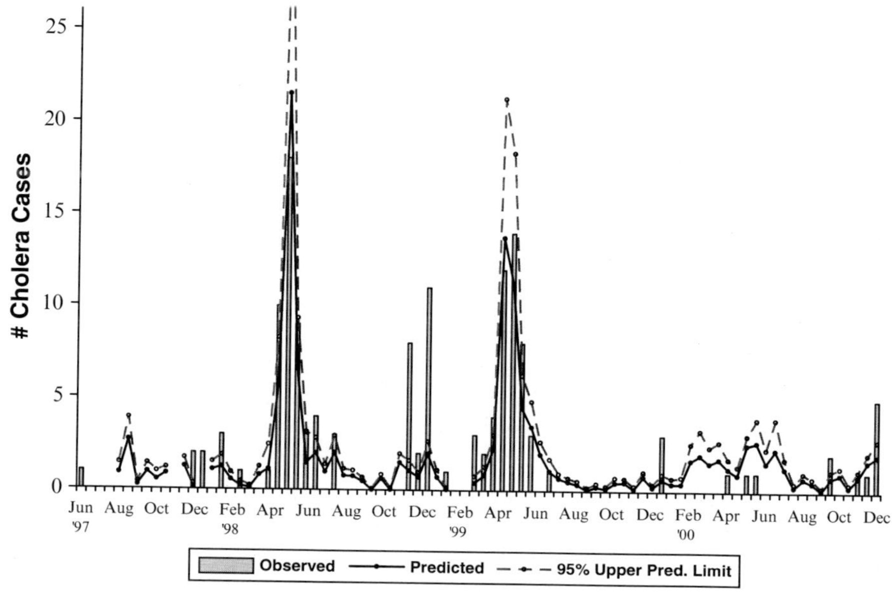

Fig. 1 Observed number of cholera cases vs. number of cases predicted by the Poisson regression model and upper 95% prediction limit when using water temperature (lag 6 weeks), *ctx*-gene probe count (lag 0 weeks), conductivity (lag 0 weeks), and rainfall (lag 8 weeks) for a lake in Bakerganj. (Reproduced with permission from the publisher: Huq et al. 2005)

Acknowledgments This work was supported by National Institutes of Health Grant No. R01A 139129-01 and National Oceanic and Atmospheric Administration Grant No. S0660009 of the Oceans and Human Health Initiative (OHHI).

References

Alam M, Sultana M, Nair GB, Siddique AK, Hasan NA, Sack RB, Sack DA, Ahmed KU, Sadique A, Watanabe H, Grim CJ, Huq A, Colwell RR (2007) Viable but nonculturable *Vibrio cholerae* 01 in biofilms in the aquatic environment and their role in cholera transmission. Proc Natl Acad Sci U S A 104(45):17801–17806

Arana I, Ortuno M, Seco C, Muela A, Barcina I (2008) Is Urografin density gradient centrifugation suitable to separate nonculturable cells from *Escherichia coli* populations? Antonie Van Leeuwenhoek 93(3):249–257

Asakura H, Ishiwa A, Arakawa E, Makino S, Okada Y, Yamamoto S, Igimi S (2006) Gene expression profile of *Vibrio cholerae* in the cold stress-induced viable but non-culturable state. Environ Microbiol 9(4):869–879

Azevedo NF, Almeida C, Cerqueira L, Dias S, Keevil CW, Vieira MJ (2007) Coccoid form of *Helicobacter pylori* as a morphological manifestation of cell adaptation to the environment. Appl Environ Microbiol 73(10):3423–3427

Blake P (1994) Endemic cholera in Australia and the United States. In: Wachsmuth I, Blake P, Olsvik O (eds) *Vibrio cholerae* and cholera: molecular to global perspectives. American Society for Microbiology, Washington, DC, pp 309–320

Boyle M, Sichel C, Fernández-Ibáñez P, Arias-Quiroz GB, Iriarte-Puña M, Mercado A, Ubomba-Jaswa E, McGuigan KG (2008) Bactericidal effect of solar water disinfection under real sunlight conditions. Appl Environ Microbiol 74(10):2997–3001

Brayton PR, Colwell RR (1987) Fluorescent antibody staining method for the enumeration of viable but nonculturable V. cholerae. J Microbiol Methods 6:309–314

Byrd JJ, Colwell RR (1990) Maintenance of plasmids pBR322 and pUC8 in nonculturable *Escherichia coli* in the marine environment. Appl Environ Microbiol 56(7):2104–2107

Cappelier JM, Besnard V, Roche SM, Velge P, Federighi M (2007) Avirulent viable but non culturable cells of *Listeria monocytogenes* need the presence of an embryo to be recovered in egg yolk and regain virulence after recovery. Vet Res 38:573–583

Chaiyanan S, Chaiyanan S, Grim C, Maugel T, Huq A, Colwell R (2007) Ultrastructure of coccoid viable but non-culturable *Vibrio cholerae*. Environ Microbiol 9(2):393–402

Chowdhury M, Xu B, Montilla R, Hasan J, Huq A, Colwell R (1995) A simplified immuno-flourescence technique for detection of viable cells of *V. cholerae* O1 and O139. J Microbiol Methods 24:165–170

Chowdhury MAR, Huq A, Xu B, Colwell RR (1997) Effect of Alum on free living and copepod associated *V. cholerae* O1 and O139. Appl Environ Microbiol 63:3323–3326

Colwell RR (1996) Global climate and infectious disease: the cholera paradigm. Science 274: 2025–2031

Colwell RR, Grimes DJ (2000) Nonculturable microorganisms in the environment. American Society for Microbiology, Washington, DC, p 354

Colwell R, Huq A (1994) Environmental reservoir of *Vibrio cholerae*. The causative agent of cholera. In: Wilson M, Levins R, Spielman A (eds) Disease in evolution: global changes and emergence of infectious diseases. Ann N Y Acad Sci 740:44–54

Colwell RR, Huq A (2001) Marine Ecosystem and Cholera. *In* The Ecology and Etiology of Newly Emerging Marine Diseases. Porter JW (ed). Kluner Academic Publisher, Dordrecht, Netherland, pp. 141–145

Colwell R, Seidler R, Kaper J, Joseph S, Garges S, Lockman H, Maneval D, Bradford H, Roberts N, Remmers R, Huq I, Huq A (1981) Occurrence of V. cholerae serotype O1 in Maryland and Louisiana estuaries. Appl Environ Microbiol 41:555–558

Colwell R, Brayton P, Grimes D, Roszak D, Huq S, Palmer L (1985) Viable but non-culturable *Vibrio cholerae* and related environmental pathogens in the environment: implications for release of genetically engineered microorganisms. Bio-Technology 3:817–820

Colwell R, Brayton P, Huq A, Tall B, Harrington P, Levine M (1996) Viable but non-culturable Vibrio cholerae O1 revert to a culturable state in the human intestine. World J Microbiol Biotechnol 12:28–31

Constanin de Magny, Guillaume, Raghu Murtugudde, Mathew RP Sapiano, Azhar Nizam, Christopher W Brown, Antonio J Busalacchi, Mohammad Yunus, Balakrish Nair G, Ana I Gil, Claudio F Lanata, John Calkins, Byomkesh Manna, Krishnan Rajendran, Mihir Kumar Bhattacharya, Anwar Huq, Bradley Sack R, Colwell RR (2008) Environmental signatures associated with cholera epidemics. Proc Nat'l Acad Sci 105(46):17676–17681

Du M, Chen J, Zhang X, Li A, Li Y, Wang Y (2007) Retention of virulence in a viable but nonculturable *Edwardsiella tarda* isolate. Appl Environ Microbiol 73(4):1349–1354

Fera MT, Maugeri TL, Gugliandolo C, La Camera E, Lentini V, Favaloro A, Bonanno D, Carbone M (2008) Induction and resuscitation of viable nonculturable *Arcobacter butzleri* cells. Appl Environ Microbiol 74(10):3266–3268

Folk R (1996) In defense of nanobacteria. Science 274:1288

Hobbie J, Daley R, Jasper J (1977) Use of nucleopore filters for counting bacteria by fluorescent microscopy. Appl Environ Microbiol 33:1225–1228

Huq A, Colwell R, Rahman R, Ali A, Chowdhury M, Parveen S, Sack D, Russek-Cohen E (1990) Detection of *Vibrio cholerae* O1 in the aquatic environment by fluorescent monoclonal antibody and culture method. Appl Environ Microbiol 56:2370–2373

Huq A, Rivera I, Colwell R (2000) Epidemiological significance of viable but nonculturable microorganisms. In: Colwell R, Grimes D (eds) Nonculturable microorganisms in the environment. ASM, Washington, DC, pp 301–323

Huq A, Sack R, Nizam A, Longini I, Nair G, Ali A, Morris G, Huda M, Siddique A, Yunus M, Albert J, Sack D, Colwell R (2005) Critical factors influencing the occurrence of *Vibrio cholerae* in the environment of Bangladesh. Appl Environ Microbiol 71:4645–4654

Knight I (2000) Molecular genetic methods for detection and identification of viable but nonculturable microorganisms. In: Colwell R, Grimes D (eds) Nonculturable microorganisms in the environment. ASM, Washington, DC, pp 77–85

Koch R (1884) An address on cholera and its bacillus. Br Med J 2:403–407, 453–459

Kogure K, Simidu U, Taga N (1979) A tentative direct microscopic method for counting living marine bacteria. Can J Microbiol 25:415–420

Kurath G, Morita Y (1983) Starvation survival and physiological studies of a marine *Pseudomonas* spp. Appl Environ Microbiol 45:1206–1211

Lee K, Ruby E (1995) Symbiotic role of the viable but nonculturable state of *Vibrio fischeri* in Hawaiian coastal seawater. Appl Environ Micbrobiol 61:278–283

Lewis K (2007) Persister cells, dormancy and infectious disease. Nature 5(1):48–56

Liu Y, Gilchrist A, Zhang J, Li X-F (2008) Detection of viable but nonculturable *Escherichia coli* O157:H7 bacteria in drinking water and river water. Appl Environ Microbiol 74(5):1502–1507

Lobitz B, Beck L, Huq A, Wood B, Fuchs G, Faruque A, Colwell R (2000) Climate and infectious disease: use of remote sensing for detection of *Vibrio cholerae* by indirect measurement. Proc Natl Acad Sci 97:1438–1443

McFeters G, LeChevallier M (2000) Chemical disinfection and injury of bacteria in water. In: Colwell R, Grimes D (eds) Nonculturable microorganisms in the environment. ASM, Washington, DC, pp 255–275

Novitsky J, Morita R (1976) Morphological characterization of small cells resulting from nutrient starvation of a psychrophilic marine vibrio. Appl Environ Microbiol 32:617–622

Nystul TG, Goldmark JP, Padilla PA, Roth MB (2003) Suspended animation in *C. elegans* requires the spindle checkpoint. Science 302:1038–1041

Pommepuy M, Utin MB, Derrien A, Gourmelon M, Colwell RR, Cormier M (1996) Retention of enteropathogenicity by viable but nonculturable *Escherichia coli* exposed to seawater and sunlight. Appl Environ Microbiol 62(12):4621–4626

Postgate J (1967) Viability, measurements, and survival of microbes under minimum stress. Adv Microb Physiol 1:1–24

Postgate J (1969) Viable counts and viability. In: Norris J, Ribbons D (eds) Methods in microbiology. Academic, London, pp 611–628

Postgate JR, Hunter JR (1962) The survival of starved bacteria. J Gen Microbiol 29:233–263

Roszak D, Grimes D, Colwell R (1984) Viable but non-recoverable stage of *Salmonella enteritidis* in aquatic systems. Can J Microbiol 30:334–338

Sack RB, Siddique AK, Longini I, Nizam A, Yunus M, Islam S, Morris JG, Ali A, Huq A, Nair GB, Qadri F, Faruque SM, Sack DA, Colwell RR (2003) A 4-year study of the epidemiology of *Vibrio cholerae* in four rural areas of Bangladesh. J Infect Dis 187:96–101

Sochard MR, Wilson DF, Austin B, Colwell RR (1979) Bacteria associated with the surface and gut of marine copepods. Appl Environ Microbiol 37:750–759

Valentine RC, Bradfield JRG (1954) The urea method for bacterial viability counts with electron microscope and its relation to other viability counting methods. J Gen Microbiol 11:349–357

Vital M, Fuchslin HP, Hammes F, Egli T (2007) Growth of Vibrio cholerae O1 Ogawa Eltor in freshwater. Microbiology 153(7):1993–2001

Xu H-S, Roberts N, Singleton FL, Attwell RW, Grimes DJ, Colwell RR (1982) Survival and viability of nonculturable *Escherichia coli* and Vibrio cholerae in the estuarine and marine environment. Microb Ecol 8:313–323

Xu HS, Roberts N, Singleton FL, Attwell RW, Grimes DJ, Colwell RR (1984) An indirect fluorescent antibody staining procedure for detection of *Vibrio cholerae* serovar O1 cells in aquatic environmental samples. J Microb Methods 2:221–231

General Model of Microbial Uncultivability

Slava S. Epstein

Contents

Abstract It has been known for over a century that only a small percent of cells from environmental samples form colonies on standard media (Great Plate Count Anomaly, Staley and Konopka (Annu Rev Microbiol 39:321–346, 1985). This chapter focuses on the causes of this disparity, and describes new cultivation technologies aiming to close the gap. It summarizes the original and literature data on the biology of "uncultivable" species is summarized, and the nature of the restrictions likely limiting the growth of these species is discussed. This analysis leads to a novel model of the microbial life cycle in nature, termed the "scout model." We argue that if microbial behavior in vivo conforms to the scout model, this will by necessity manifest itself in vitro as the Great Plate Count Anomaly. The scout model also draws connections to other aspects of microbial behavior, such as viability – but not cultivability – of some cells, an apparent slow growth of certain species, seeming ability of microbes to persist in the presence of unfavorable factors, including antibiotics, and latent infections.

S.S. Epstein
Department of Biology, Northeastern University, Boston, MA, 02115, USA
e-mail: s.epstein@neu.edu

Microbiol Monogr, doi:10.1007/7171_2009_2

1 Introduction

A great part of natural microbial diversity is missing from our laboratories (Staley and Konopka 1985; Rappe and Giovannoni 2003; Schloss and Handelsman 2004). The question is, why? Have our previous efforts been lacking in discipline, or is there something critical missing from the traditional approach altogether? If we were not limited by practical considerations, and could take all the media ingredients present in our labs, mix them in all possible combinations, match that with an equally high diversity of incubation conditions, and in doing so exceed the totality of cultivation experiments performed over the past century, – would that close the gap between existing and cultivated microbial diversity? This might appear a silly question, because undoubtedly such an enterprise would result in a score of new cultures. But, would the overall *percent* recovery begin to approach 100%? A positive result would indicate that we only needed to roll up our sleeves. If the change was nominal, it would be reasonable to think that microbiologists have utterly overlooked some principal aspect(s) of microbial life style, and that identifying these aspects should be a priority before replicating the cultivation approaches of the past.

This work attempts to distinguish between these two scenarios, and considers possible alternatives. It describes the cultivation approaches we have developed to access "uncultivable" microorganisms (Sect. 2), a path to their domestication (Sect. 3), and the resolution of what constitutes uncultivability in at least one model environmental species (Sect. 4). The work proceeds with describing single-cell cultivation experiments, and what we have learned from them about uncultivables (Sect. 5). This information is then synthesized in the form of a recently proposed model (Epstein 2009) model that attempts to explain the Great Plate Count Anomaly as a consequence and manifestation of microbial growth strategies in nature (Sect. 6). Section 6 also discusses possible implications of the model for microbial cultivation methodology, as well as its connection and relevance to other microbiological concepts and phenomena, such as viable but nonculturable (VBNC) cells and latent infections, among others.

Throughout this work, we will repeatedly use the term "uncultivable" microorganisms. This term is widely used, but is to some extent misleading. Since uncultivables are able to grow in nature, they are not really uncultivable but rather "as yet not cultivated." The term should be appended as "presently uncultivable in the laboratory under available culture conditions." It is in this sense that we will use the term uncultivable, choosing it exclusively for the sake of brevity.

2 In situ Cultivation of Uncultivable Microorganisms

Is it possible to grow something, e.g., a microbial species, that has unknown growth requirements, without first resolving these unknowns? The answer is likely "yes," because, assuming all species grow in nature, their cultivation might be possible via a reasonable imitation of the natural conditions conducive of this species' multiplication.

Note that in principle the natural milieu can serve as an ultimate growth medium even if we do not know all its components. Considering microbial cultivation from this point-of-view suggests that, curiously, cultivation per se does not seem to be a problem: if a cell from a growing population is removed from its natural habitat, and then – instead of going to Petri dish – is returned back to the environment, chances are it will grow again. This growth could be equated with cultivation, except of course the results of such "cultivation" are inaccessible, for the newly cultivated species is (again) mixed with all its neighbors. Nonetheless, this suggests that the real problem is not necessarily how to grow environmental microorganisms – once more, this is already happening in their own environment – but rather how to contain this growth. We proposed to achieve such containment by placing environmental microorganisms inside diffusion chambers incubated in the natural habitat of these organisms (Kaeberlein et al. 2002).

2.1 Method 1: Diffusion Chamber

The first diffusion chamber we designed and used is a simple metal washer sandwiched between two polycarbonate membranes with ≈0.03-μm pores (Fig. 1). The inner space is filled with agar mixed with microorganisms from a test habitat; membranes separate these microorganisms from the environment thus providing containment. Upon the return of this assembly to the natural habitat (or its simulated

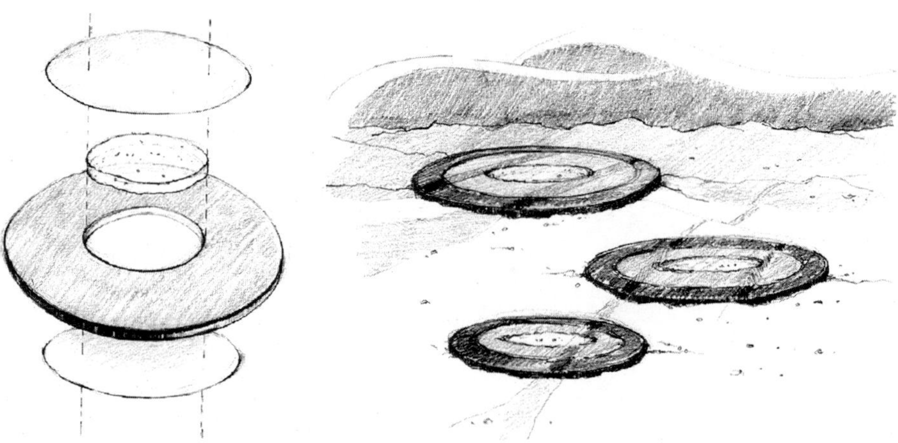

Fig. 1 Diffusion chamber for in situ cultivation of environmental microorganisms. The *left panel* shows the schematics of the chamber assembly: a membrane is glued to the bottom of a metal disk with central orifice; agar mixed with test cells is poured into the membrane forming an agar slab with embedded cells; the second membrane is glued to the upper part of the disk, sealing the inner space. The assembly is then returned to the natural environment of the test cells, as shown on the *right panel*. Artwork by Stacie Bumgarner, Whitehead Institute for Biomedical Sciences, Cambridge, MA, USA

version, such as a large aquarium filled with freshly collected soil, aquatic sediment, etc.), diffusion through pores establishes chemical communication between the cells and nature. The expectation is that the cells will behave in a fashion similar to their "free-living" counterparts: if a species grows in nature at the time of the experiment, it should grow inside the chamber as well. These expectations were met in a proof-of-concept growth experiment, which showed a 300-fold higher recovery rate of marine sediment microorganisms than from parallel experiments that utilized more traditional cultivation techniques (Kaeberlein et al. 2002). Further experimentation with diffusion chambers produced a score of species whose growth could be easily maintained once incubated in nature, but failed to grow on standard media in the lab (Fig. 2). This suggested that, apart from a large number of cells forming colonies inside the diffusion chamber, the species composition of the grown material was also markedly different between the chamber and standard Petri dish. Later on, we obtained experimental evidence supporting the claim (Bollmann et al. 2007). The chamber-reared microbial cultures became our model systems for exploring the nature of growth limitations imposed by the in vitro environment (see Sect. 4).

The diffusion chamber-based approach represents a significant departure from traditional cultivation. It does not require knowing – or guessing – about the biology

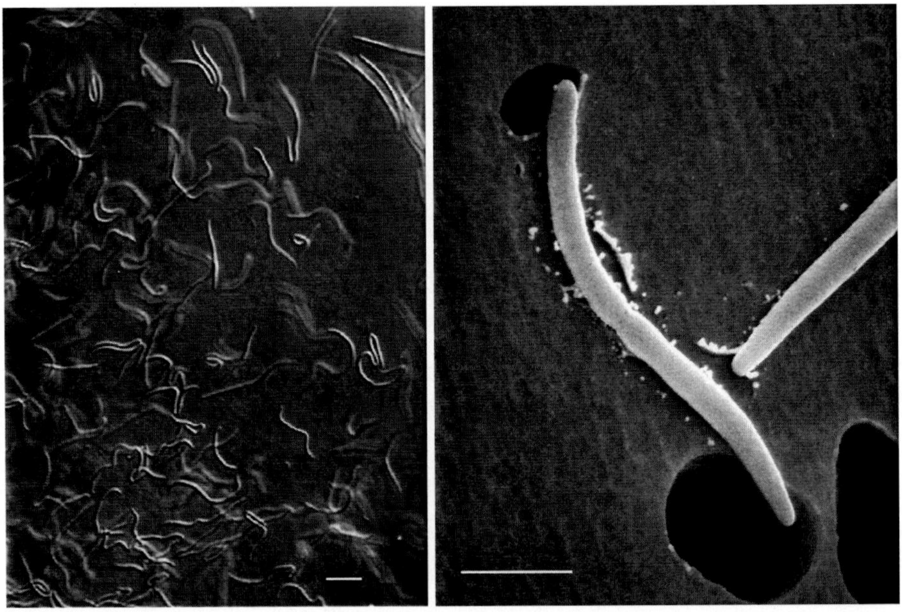

Fig. 2 Uncultivable microorganism MSC2 sharing 97% 16S similarity to *Arcobacter nitrofigilis*: differential interference contrast (*left panel*) and SEM (*right panel*) views. MSC2 was isolated by Kaeberlein et al. (2002); it could be maintained in diffusion chambers in vivo but did not grow in isolation in vitro

of target species, and it makes growth media manipulations largely unnecessary. This approach relies on a simple observation that nature by definition contains everything that microorganisms inhabiting it require for growth – and makes use of these growth components, whether they are known or not. The success of this approach was confirmed by later studies that were based on a similar idea. For example, Ferrari et al. (2005) used a single membrane placed between soil microorganisms and a soil sample to provide the cells with soil constituents through diffusion. The growth of cells belonging to, for example, the elusive unculturable bacterial division TM7 was observed. Aoi et al. (submitted) inoculated aquatic cells into hollow fiber tubes with semipermeable walls, bathed the tubes with the source water, and observed microbial recovery exceeding that typically registered by in vitro studies.

The use of these devices has certain limitations. Single membranes do not form an enclosed space and may be prone to contamination. The hollow fiber tubes can be used to cultivate planktonic cells but their application for sediments and soil may be problematic because they may be too large to imitate sharp, sub mm–scaled natural chemical gradients. The diffusion chamber in Fig. 1 appears to be largely free of these limitations, but it does share with other devices an even more critical complication. Environmental cells placed into the chamber and incubated in situ typically form very small colonies invisible to a naked eye; as a rule, these do not grow in size even after long incubations. The same was observed by others (Ferrari et al. 2005). Visualization of such colonies for counting, and especially their sampling for subculturing, requires a high power compound microscope. This makes detection and isolation into pure culture tedious, and high-throughput cultivation nearly impossible. To resolve these bottlenecks, we designed a variant of the diffusion chamber, hereafter referred to as "the isolation chip," or ichip for short, that greatly facilitated diffusion chamber-based microbial cultivation and enabled large-scale isolation.

2.2 Method 2: Ichip

One way to resolve logistical difficulties in detecting and sampling microcolonies is to miniaturize diffusion chambers and inoculate them with single cells. The first modification allows for quick scanning of the contents of multiple chambers, whereas the second provides means for growing pure cultures in one step. A practical solution to both was suggested by the OpenArray™ technology developed by BioTrove, Inc. and designed for the precise capture of small samples (http://www.biotrove.com/global/pdf/SPIE_2001.pdf). The OpenArray is a flat plate with multiple through-holes. Proprietary processes are used to coat the surface of the plates so that they are hydrophobic, while rendering the interiors hydrophilic and biocompatible. When such a plate is dipped into a cell suspension, each hole captures a certain volume of suspension containing a certain number of cells. When the suspension is appropriately diluted, this number may be on average one (cell). If the suspension is an agar-based medium, then upon solidification the cells will be immobilized inside small agar plugs and become "trapped" in their respective through-holes (and thus isolated from each other). Sandwiching the plate between semipermeable barriers, such as 0.03-μm pore size polycarbonate filters used in the original diffusion chamber, effectively transforms the loaded plate into an array of many diffusion minichambers (the ichip). Its in situ incubation provides the cells inside with their naturally occurring growth components. Cells are expected to grow and form colonies at least as well as they would in "standard" diffusion chambers described in the section above. If only a single cell is present in each minichamber, its growth will automatically produce a pure culture. In this way, cultivation and isolation of uncultivables can be achieved in large numbers and in one step.

Presently, we are using a variant of the ichip, shown in Fig. 3, for large-scale isolation of microorganisms from soil and aquatic environments. This variant is built entirely from hydrophobic plastic polyoxymethylene, commonly known under DuPont's brand name Delrin®. It contains 374 through-holes per assembly, arranged in two identical arrays; of course, any other configuration is possible, including one where the holes match wells of a microtiter plate. The holes in this specific configuration capture about 1 μl of cell suspension. By using a cell suspension containing one cell in 1–10 μl, we achieve loading of one cell into 1–10 holes, virtually ensuring that no agar plug contain more than one cell. The through-holes in each array are positioned so that one standard, 25-mm in diameter, polycarbonate filter fully covers the array on one side; four filters are needed for one assembly to separate all its through-holes from the outside environment. We observe that the ichip affords

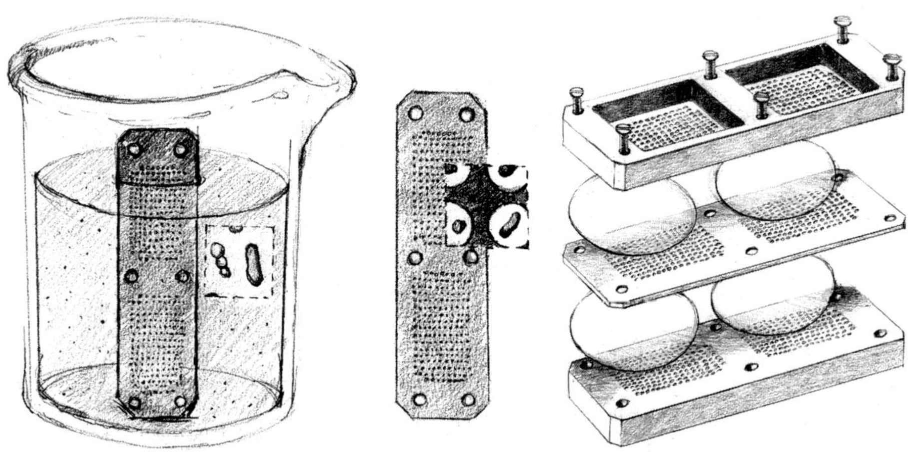

Fig. 3 Isolation chip, or ichip, for high-throughput microbial cultivation in situ. *Left panel* shows dipping a plate with multiple through-holes into a suspension of cells such that each hole captures (on average) a single cell, as illustrated in the central panel. *Right panel* shows how ichip is assembled: membranes cover arrays of through-holes from each side; upper and bottom plates with matching holes press the membranes against the central (loaded) plate. Screws provide sufficient pressure to seal the content of individual through-holes, which become miniature diffusion chambers containing single cells. Artwork by Stacie Bumgarner, Whitehead Institute for Biomedical Sciences, Cambridge, MA, USA

better microbial recovery than the "standard" diffusion chamber (Fig. 1). This may be a result of (1) a shorter diffusion path between the cell in the ichip and the outside environment, so that more cells respond to their environments and grow, as well as (2) the ease at which growth of such cells can be visualized under a microscope, so that fewer colonies are missed. Removal of grown colonies is easily achieved by touching the agar plug with a sterile toothpick or paper clip, making the plug fall into, for example, a well of a microtiter plate. The combination of these factors makes the ichip suitable for large-scale cultivation and isolation of environmental microorganisms: a parallel incubation of several ichips provides dozens to hundreds of pure (and mostly novel) microbial cultures.

3 In vitro Growth of Uncultivable Microorganisms and their Domestication

3.1 Raising Cultivable Variants of Uncultivable Species

In our working definition, an uncultivable is a microorganism that does not grow on standard media in vitro but forms colonies in diffusion chambers in situ (and is therefore viable, not damaged, not an exceedingly slow grower, etc.). Working with such uncultivable isolates we noticed that their multiple transfers through the chamber typically led to the appearance of cultivable variants capable of sustainable growth in the lab (Fig. 4; Nichols et al. 2008). A qualitative illustration of this is the microbial richness we recovered from the apparently lifeless sands collected in Dry Valley, Antarctica: once cells from these sands went through a growth cycle in a diffusion chamber incubated within the sand sample, they produced a glorious diversity of numerous cultivable isolates (Fig. 5).

Further quantitative experiments suggested these observations were not accidental. In a separate study of uncultivables from a fresh pond environment, we passaged a sample through several rounds of cultivation in diffusion chambers, determining at each step what species from the grown material could be cultivated in a standard Petri dish. This gave us a collection of domesticated strains (Bollmann et al. 2007). We compared this collection with species cultivated using the traditional approach, when an environmental sample is directly inoculated into a Petri dish, without the use of a chamber. The difference between the two collections was substantial. The isolates unique to the diffusion chamber approach outnumbered the "traditional" isolates by a factor of 3, with only 7% of the total 438 isolates shared between the collections. More importantly, the domesticated pool included representatives of rarely cultivated phyla, such as *Deltaproteobacteria*, *Verrucomicrobia*, *Spirochaetes*, and *Acidobacteria*, whereas the traditional pool did not. It is worth noting that some of the more interesting isolates appeared to have acquired the ability to grow in vitro only after multiple (up to 3) transfers through the diffusion chamber. In a follow-up study of *Actinobacteria*, we again observed a very substantial

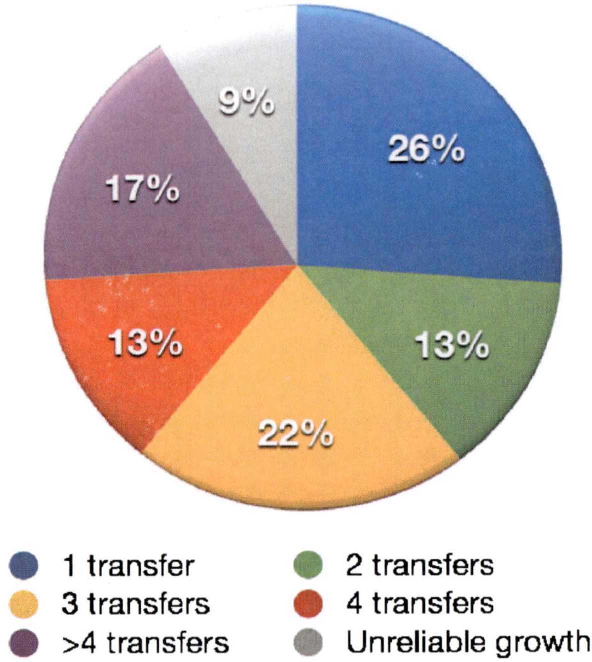

Fig. 4 Percent of uncultivable isolates producing domesticated variants after transferring their colonies through a series of incubations in diffusion chambers; original data from Nichols et al. (2008)

Fig. 5 Stereomicroscope view of domesticated microorganisms from Dry Valley sands, Antarctica. This diversity of cultivated strains was achieved once environmental cells were first grown in a diffusion chamber incubated within a Dry Valley sand sample. The sample is courtesy of Sam Bowser, Wadsworth Center, NYS Department of Health, Albany, NY, USA

number of strains appearing on Petri dishes only after rounds of cultivation in situ (Bollmann et al. unpublished results). This strongly suggests that in situ incubation facilitates the appearance of cells that have fewer growth restrictions, and are capable of growing in vitro. Though the exact nature of the domestication process remains somewhat mysterious, in practice, it provides convenient access to a substantial pool of novel microbial diversity. For example, after we used in situ incubation to grow environmental microorganisms from the Oak Ridge National Laboratory field sites in Tennessee, we discovered representatives of the Genus *Frateuria* (Bollmann et al. submitted) and other species among the domesticated species, previously known from the sites exclusively from a culture-independent metagenomics program (Zhou, personal communication).

3.2 Symbiotic Growth of Microorganisms and Coculture

We discovered that our uncultivable isolates could also be grown in the Petri dish by an alternative method: in coculture with selected other microorganisms. One example of such synergistic growth involves some of our first diffusion chamber-reared isolates, MSC1 (93% 16S rRNA sequence similarity to its closest relative *Lewinella persica* of the Cytophaga–Flexibacter–Bacterioides group) and MSC2 (*Arcobacter* relative; Fig. 2) (Kaeberlein et al. 2002). While we could maintain their pure cultures exclusively inside diffusion chambers, serendipitous mixing of the two showed their ability to grow together on standard media. The pattern of colonies in a Petri dish appeared to show codependence: denser colonies of MSC1 formed a gradient of increasing size converging on diffuse colonies of MSC2 (Fig. 2d in Kaeberlein et al. 2002). We have since tested a number of uncultivable isolates against a panel of cultivable environmental species from our culture collection and found an appropriate "helper" for all tested uncultivables (Nichols et al. 2008).

It is possible that the observed growth synergy is based on cross-feeding. This explanation is a natural null hypothesis since there are numerous examples of beneficial, joint use of resources by members of microbial consortia (Schink 2002; McInerney et al. 2008). However, all our cocultures were observed on rich media (technical grade casein, marine broth, LB), a possible – but unlikely – environment for cross-feeding. Further evidence against cross-feeding comes from our experimentation with one of the uncultivable isolates obtained via the diffusion chamber approach, MSC33, which shared 95% 16S rRNA gene sequence identity with *Psychrobacter* sp. (Nichols et al. 2008). As was the case with other uncultivables, we were able to raise its cultivable variant, MSC33c. The cells of MSC33c formed clearly visible macrocolonies on the same agar media on which the cells of MSC33 refused to grow. We then paired the two strains in a Petri dish and observed the induction of growth of MSC33 by MSC33c. Several rounds of cocultivation of a population of MSC33 with MSC33c led to the appearance in the former population of cells capable of self-sustained growth. We thus observed the birth of domesticated cells in an uncultivable population in the presence of helpers, in this case domesticated cells of the same strain. The study of MSC33 indicated that domestication by repetitive incubation in diffusion chambers and induction of growth in the presence of a helper may be related: the former results in domesticated cells, which then act as helpers in the latter. Notably, MSC33 could also be induced to grow in artificial media by the presence of other, unrelated helpers, such as MSC 105, an isolate sharing 99% 16S rRNA gene similarity to *Cellulophaga lytica.* The model we introduce later in this work will account for these observations; here we would like to note only that both domestication and cocultivation with a helper may be general methods for growing uncultivables on standard media in the lab.

4 Mechanism of Uncultivability: Case Study of MSC33

The empirical observations described in Sects. 3.1–3.2 are difficult to accommodate by any conventional theory of the nature of microbial uncultivability (e.g., media are toxic and/or imperfect; cells are damaged, stressed, or slow growing, etc.). In a search for a plausible interpretation of our findings, we hypothesized that at least some uncultivable microorganisms require specific signals originating from their neighbors (Kaeberlein et al. 2002), including their growing kin (Nichols et al. 2008). This implies that since signals trigger cell growth, the absence of such signals from standard media may at least in part explain the Great Plate Count Anomaly. Below we provide a summary of the experimental support for this signaling hypothesis (Nichols et al. 2008).

We chose the closely related strains MSC33 and MSC33c as our model organisms and aimed to dissect the chemistry of their interactions. We discovered that MSC33c spent medium preserved the growth stimulating activity. This simplified the natural product chemistry aspect of this research. Bioassay-guided fractionation of MSC33c spent medium coupled with LC-MS eventually led to identification of a 5-aminoacid peptide LQPEV that, at 3.5 nM concentration, replicated the effect of MSC33c presence and induced colony formation in MSC33. A simple addition of the synthetic version of this peptide to standard media transformed an uncultivable MSC33 into a microorganism that grows well in vitro. At such low concentrations (several orders of magnitude lower than typical oligotrophic conditions, Kuznetsov et al. 1979), the peptide was an unlikely energy or carbon source. Instead, the pattern of its activity pointed towards its role as a signal, and thus the signal-based growth regulation of MSC33 (and possibly other species as well). This signaling hypothesis is in line with earlier observations of increased microbial recovery in the presence of cyclic-AMP and *N*-acyl homoserine lactones (Bruns et al. 2002).

The growth of at least some uncultivable species thus appears to be one of many microbial processes regulated by cell–cell communication (Waters and Bassler 2005; Camilli and Bassler 2006). The fact that uncultivables respond to the presence of their growing kin suggests that, in a natural environment, the mere presence of nutrients may not be sufficient to initiate cell division. After all, in the environment a sudden availability of nutrients may be a short-lived random fluctuation, and initiating growth in response to their ephemeral presence may be risky. In contrast, signals coming from *multiplying* kin are indicative that the environment has *for a period of time* been conducive to growth. The fact that uncultivables respond to the presence of selected unrelated species suggests that in the environment, nongrowing cells may be waiting for cues from their, for example, syntrophic partners. This would ensure the growth of a species only when and where its partners are already present in sufficient numbers.

These considerations make it very tempting to explain a significant part of the Great Plate Count Anomaly by the failure of standard media and traditional cultivation techniques to provide microorganisms with growth-promoting signals. While we believe signal-based growth regulation is indeed such a partial explanation, we would like to point out that, in fact, the very observations that lead to this explanation contain a paradox of significant importance. Indeed, if the cells whose growth is triggered by signaling compounds can produce domesticated variants that are independent of such signals, and if they do so as easily as we observed (Sect. 3.1, Fig. 4), then such cultivable variants are expected to be part of the natural uncultivable populations as well. If so, why is it that we do not see such variants growing in Petri dishes? In other words, why have we not cultivated the majority of environmental *species* – even if the majority of their *cells* are signal-dependent? As the next section describes, the story of at least some uncultivable species may be even more intriguing than an explanation relying on "just" the signal-based regulation of their growth. The next section will present several relevant findings, which, as often happens, started from observations seemingly unrelated to the question at hand.

5 Heterogeneity in Microbial Populations

Observations described in Sects. 3 and 4 suggest that at least some uncultivable populations contain a certain percentage of cells that are different from the rest of the population in that they lack growth restrictions typical of the population's majority. Because these cells are capable of growing in vitro, and from the point-of-view of a microbiologist they are essentially domesticated, these cells are a valuable resource in the search for novel species. The question is how do we find such cells and the colonies they produce, considering that they may not be very abundant?

The suspicion has been that quickly growing "weed" species could prohibit – or mask – in vitro growth of other, perfectly cultivable species that are either rare, do not grow as quickly, or both. The ultimate tool to eliminate the influence of such weeds (and, for that matter, of any other species), would be to separate the inoculum into single cells, and plate each cell in its own miniature Petri dish. This is similar to cultivation by dilution to extinction, known to increase microbial recovery (Button et al. 1993; Connon and Giovannoni 2002), and sometimes produce spectacular isolates (Rappe et al. 2002). The single cell approach gives every cultivable cell a chance to grow. This also allows for long-term cultivation experiments thought to facilitate recovery of novel (slow growing) microbial diversity (Davis et al. 2005). We realized that an analogue of the central plate of an ichip (Fig. 3) may be a practical tool for separating large numbers of cells for single-cell cultivation. In an exploratory fashion, we diluted marine sediment microorganisms in agar supplemented with 0.1% v/w Luria Broth. When a plate was dipped into the suspension, each through-hole captured on average a single cell. In this experiment, we captured 5,000 cells into 5,000 through-holes; these holes, upon solidification of the agar, became miniature Petri dishes for single cell cultivation. We incubated the plates for >4 months in a humidified chamber, periodically examined them under a dissecting microscope, counted agar plugs showing growth, and removed them for microbial identification (Buerger et al. 2008). This simple experiment led to two unexpected findings.

First, a surprisingly large number of microbial colonies did not appear until after 1 month of incubation. The cumulative recovery showed little sign of leveling off during the second month of the experiment, and showed a moderate decline over the following 2 months (Fig. 6). Curiously, the majority of the cultures obtained after 3–4 months of incubation did not appear to be inherently slow growing: when subcultured, 84% of such isolates formed visible growth after just 48 h. Therefore, the majority of apparent slow growers may in fact be capable of fast expansion – as shown by the speed of their regrowth.

Second, species detected multiple times in this experiment did not come as a single batch of cultures but appeared gradually over the 4 months of incubation, thus showing no species-specific lag period. For example, *Pseudomonas argentinensis* was registered at all time points. This means that some cells of *P. argentinensis* grew within the first few days of incubations, others grew within several weeks, and yet another category of cells of *P. argentinensis* did not grow until the very end of

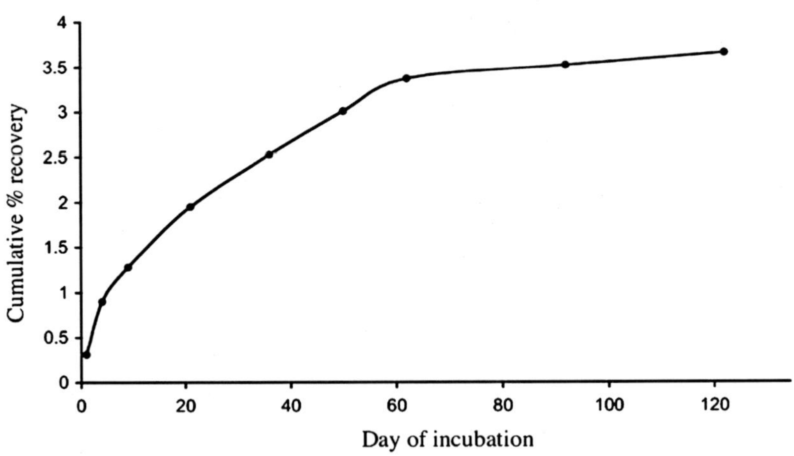

Fig. 6 Microbial recovery during long-term incubation. Microbial cells from marine sediments were incubated individually in miniature Petri dishes

the experiment. A tentative conclusion is that *P. argentinensis*, and selected other populations, are neither fast nor slow growing but rather heterogeneous with respect to the growth preferences of individual cells.

The overall conclusion drawn from this preliminary experiment is that either the cells in a microbial population grow at dramatically different rates, or else they grow at more or less the same rate but exhibit drastically different lag phases. This heterogeneity in environmental populations is consistent with the idea of microbial cell individuality, which has been abundantly shown even for members of the same clonal population (Avery 2006; Dubnau and Losick 2006). Note that heterogeneity in a population can only manifest itself in an experiment involving sufficient number of cells, and this may have important implications for how results of microbial cultivation experiments should be interpreted, e.g., in our experiment we manipulated a mix of 5,000 environmental cells, among which some species must have been represented by just one to several cells. If *P. argentinensis* in our experiment was also represented by a few cells, it would be possible for its colonies to appear more or less simultaneously simply by chance. If this happened at the beginning of the experiment, *P. argentinensis* would appear fast growing; if this happened at the end of the experiment, it would appear a slow grower; – while from the actual experiment we learned it might be neither. It follows that when a colony is detected in the Petri dish only after prolonged incubation, it does not necessarily mean it was produced by an inherently slow-growing species.

6 Explaining the Great Plate Count Anomaly

The section attempts to explain the experience gained in cultivating microbial species. The Great Plate Count Anomaly states that there is a significant disparity between the number of cells in the sample and the number of colonies they form in a Petri dish. We recognize that a significant portion of a sample's diversity may be missing from the dish because some of the trivial nutrients might be present at suboptimal concentrations, or their specific mix might be toxic, or some more exotic – but required – nutrients might be absent. We argue however that even a combination of these reasons cannot fully account for the phenomenon, as some species exhibit growth patterns incompatible with the above explanations (illustrations of such patterns are given in the above sections). Here we detail a recently introduced model (Epstein 2009) that attempts to explain and account for the following "anomalies" in growth behavior: (1) uncultivable microorganisms form cultivable variants; (2) these domesticated variants can induce the parental uncultivable population to grow (in vitro); (3) similar induction is observed in the presence of unrelated helper species; (4) induction appears to be based on growth-promoting signals; (5) populations appear to be very heterogeneous with respect to growth preferences of its cells: after the addition of nutrients, some cells start dividing quickly, others wait for weeks or months before showing growth; and (6) populations that initially took months to grow, after subculturing may grow within days, and sometimes hours.

6.1 Model

We propose that microbial behavior in a Petri dish is an attempt by microorganisms to replicate in vitro their in vivo life styles and cycles. One of the best known and researched microbial cycles is spore formation, as a survival mechanism, and germination, as an exit strategy. We recently suggested (Epstein 2009) that non-spore-forming species achieve essentially the same goal of surviving under hostile conditions, and multiplying under favorable ones, by cycling between dormancy and activity, though not necessarily by forming morphologically unique specialized structures. The proposed model describes three distinct strategies for such cycling and argues that a mix of these in vivo strategies would by necessity manifest itself in vitro as the Great Plate Count Anomaly. We first provide a description of the model, followed by an analysis and rationalization of its key assumptions.

The first strategy is the simplest one; we call it "the scout strategy" (Fig. 7a). According to the model, adverse environmental conditions cause microbial cells to enter a dormancy state; in fact, most microbial cells in nature might be in this state most of the time. There is little doubt that inactivity allows for a prolonged survival; this has been proposed repeatedly over the past century (see below for more on the subject). The novelty of the model is in hypothesizing how microbes exit dormancy, namely via cells termed "scouts."

A scout is an active cell produced by the silent, dormant cells. Scout formation is proposed to be a result of noise-driven, stochastic, low-frequency events in dormant cells, such as a stochastic change in the expression or repression of a master regulatory gene. Therefore, the scout is neither a genetic variant nor a specialized cell. Indeed, the scout may be identical to a typical cell in an actively growing population.

The term scout comes from its function: the scout explores resources available at the moment. If adverse conditions persist, the scout dies after it has exhausted its internal resources, but is then replaced by a succession of new ones, stochastically formed within the same population. If a scout forms under growth-permissive conditions, it will start a new population, thus achieving the main objective: multiplication of its genome (Fig. 7a). Note that less abundant populations may lack scouts due to chance. Under the scout model, these populations, upon plating in a Petri dish, will likely appear uncultivable regardless of the growth conditions offered.

The second scenario (Fig. 7b) adds certain sophistication to the scout strategy: the scout cell with signaling function. We propose that at least some species produce growth-promoting signals (see also Nichols et al. 2008). When the scout finds conditions conducive of growth, it starts a new population. If each cell in the population produces a certain amount of growth-inducing signal(s), population growth will lead to accumulation of the signal. In a quorum-sensing (Fuqua et al. 1994) fashion, this will eventually wake up at least a portion of the original – and up to this point still dormant – population. In this scenario, a cell can switch from dormancy to activity in two ways: first, as a result of noise in the gene network (scout), and second, as induced by a signaling compound. Some strains may require such compounds only for activation; others may be dependent on this compound's

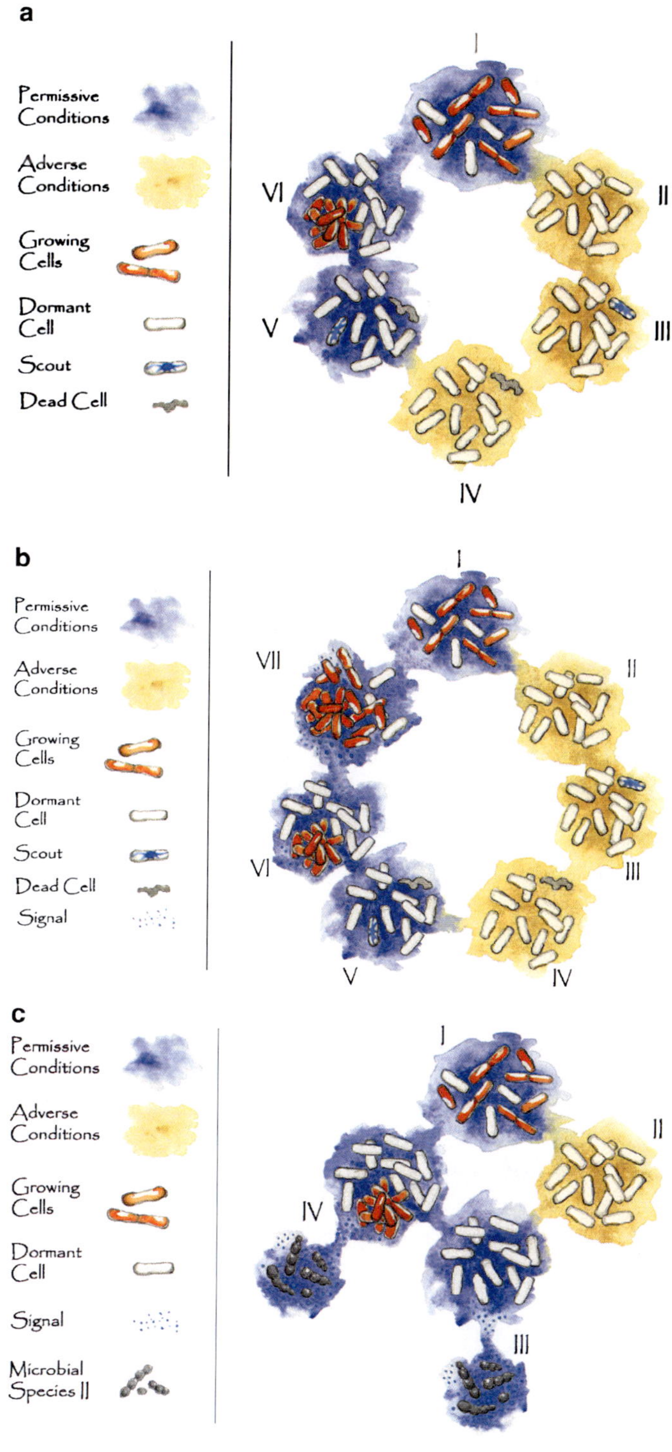

Fig. 7 Microbial strategies of survival in nature: dormancy under adverse conditions, growth under favorable conditions, with transition via scout cell and/or growth-promoting signaling compounds. *Explanations in the text*. Artwork by Stacie Bumgarner, Whitehead Institute for Biomedical Sciences, Cambridge, MA, USA

sufficient concentration during the entire active stage. Regardless, the central element of this strategy is still the scout cell, but now with an additional signaling function. Whether such a population appears cultivable or uncultivable in vitro is still determined by a chance event – the presence or absence of scouts.

The third strategy (Fig. 7c) is also based on cell–cell signaling but proposes a different source for the signal: an unrelated helper species. According to this model, when microbial species grow, the signals they produce induce dormant populations of certain neighboring species to start growing as well. Populations employing a strategy of waiting for a signal from helper strains are unlikely to grow in vitro in isolation, unless they are also capable of one of the two strategies discussed earlier.

The proposed model thus rests on three postulates: microbial cells can enter dormancy; the exit from dormancy can be a stochastic result of noise in gene networks; and/or dormant cells can be induced to grow by signaling compounds. Today all three processes remain hypothetical, but the currently available data support their reality.

Dormancy. The idea that nonspore-forming species can enter a state of suspended animation has a long history. One of the earliest notions of inactivity comes from works of Sergei Winogradsky (1924). He introduced the category of zymogenous, or transient, species, by implication mostly dormant. For half a century following that publication, inactivity was hypothesized numerous times to explain microbial survival, often under different names, such as dormancy, resting stage, pseudosenescence, moribundity, etc. (cf. Butkevich and Butkevich 1936; Jannasch 1959; Gray 1976; Postgate 1976).

During that period the idea of dormancy was viewed by many as anathema, but eventually the weight of empirical evidence – albeit indirect – in favor of some form of resting stage prevailed. In 1978 the idea of microbial dormancy was presented as a formal hypothesis (Stevenson 1978). Soon after, ultramicrobacteria were discovered in the environment; their small cell size was hypothesized to be the phenotype of dormant cells (Novitsky and Morita 1976, 1978; Torrella and Morita 1981).

In another development, dormancy became the cornerstone of the "viable but nonculturable" concept, proposed a quarter of a century ago by Colwell's group to explain survival of enteric bacteria in the natural environment (Xu et al. 1982, see also Colwell, "Viable but not Cultivable Bacteria" in this volume). VBNC state is a hypothetical condition in which the cell is undetectable by standard cultivation yet remains viable (and, in the case of pathogens, capable of initiating infection). The mechanisms of entering VBNC state, maintaining it, and eventual cell revival have not been explained or proposed, and its very existence remains controversial (see Sect. 6.3.2). Nonetheless, helped by a significant amount of indirect evidence, the idea of a VBNC state has reemphasized the potential importance of microbial dormancy.

No longer controversial, the dormancy state is now widely used to explain tolerance of cells to otherwise challenging and extreme factors (see Lewis, "Persisters, Biofilms, and the Problem of Cultivability" in this volume). However, the mechanisms for entering and maintaining dormancy are still unknown. Strictly speaking, the very existence of dormancy remains unproven. If microbial dormancy exists, why is it so difficult to irrefutably demonstrate it? This is likely due to a very peculiar challenge: the researchers must distinguish between cells that are dormant, and cells that are simply dead. The ultimate test that distinguishes between the two is of course cell revival, but its demonstration requires having a dormant population. Therefore, the researcher must have a way of telling the difference between dormant and dead cells prior to proving this difference exists … which is not unlike the proverbial Catch 22. In principle, it is possible that efforts to revive allegedly dormant cells largely fail to do so because they are futilely applied to cells that have been dead all along. One wonders if there is a difference between a genuinely dead cell and a viable but dormant cell that remains indefinitely so because we do not know how to revive it. In the microbial world, the difference between life and death may be a philosophical one.

However, if an inactive state does in fact exist, some of the past revival experiments should have employed – whether by design or chance – truly dormant cells. If so, why did we not witness cell reactivation in those cases? After all, microbiologists did learn how to revive bacterial spores (stress leads to their germination), and it is rather trivial to show how resting plants and hibernating animals come back to life (in response to environmental cues). Why is the case of nonspore-forming microbial species any different? The answer may be simple: the revival has in fact been observed, but has not been interpreted as such. Dormant nonsporulating microorganisms may have a mechanism of revival that is unlike any of the above. Not conforming to our experiences and behaving contrary to our expectations, the "abnormal" pattern of their revival may be easy to overlook and easy to ignore. The scout model proposes one such pattern: stochastic exit, as described below.

Stochastic exit from dormancy. Since most aspects of dormancy are hypothetical, the mechanism of existing dormancy is by necessity a matter of speculation. That said, random conversion of one phenotype (dormant cells) into another (active cells, or scouts) proposed here, parallels several known processes that subdivide clonal populations into two contrasting phenotypes. Phase variation is one example of stochastic and reversible switches between the expressions of different alleles of one gene (Zieg et al. 1977) (note however that this is actually a genetic change). Genetic competence in *Bacillus subtilis* provides an example of epigenetic, noninheritable, stochastically produced heterogeneity in clonal populations, in this case in the cells' ability to take DNA from the environment and recombine it with its own DNA. The advent of methods to study single cells made it abundantly clear that cells' heterogeneity with respect to their gene expression pattern is a general phenomenon, resulting in a concept of microbial bistability (Maamar and Dubnau 2005; see also Ferrell 2002; Avery 2006; Dubnau and Losick 2006). Bistability can arise from a number of mechanisms, including positive autoregulation of a gene, such that minimal fluctuations in its activity can cause a spiraling effect on the downstream genes (and thus heterogeneity in the population), or a system of two mutual repressors, that has the same effect (Gardner et al. 2000; Becskei et al. 2001). Further examples of stochastic events triggering switching of cell phenotype was provided by Choi et al. (2008). Regardless of the specific mechanism at work, the result is a population of isogenic cells represented by two phenotypes, which quite possibly provides a selective advantage in a fluctuating environment

(Avery 2006; Dubnau and Losick 2006). That such bifurcation can happen in response to environmental stress (e.g., *B. subtilis* competence) further indicates the environmental relevance of the phenomenon. We hypothesize that a similar (bifurcation) process creates and maintains two kinds of cells in a clonal dormant population: the dormant majority and the active minority (scouts). One intriguing possibility is that the proportion of the latter may be under selective pressure, so that the amount of "noise" responsible for scout formation has been selected in the course of evolution to activate a specific percent of dormant cells over a given period or time. Regardless of the mechanism, the proposed process effectively wakes up the affected cell, which to an outside observer would appear as a random (and perhaps rare) event.

Signal induction of growth. Presently, there is no direct proof that growth of a dormant cell can be induced by signaling compound(s). However, there are numerous indications that the process operates in at least some microbial species. In Sect. 4, we described the growth behavior of our model uncultivable microorganism, MSC33, which starts growing in artificial media in the presence of nanomolar concentrations of specific short peptides (but not their individual constituents) (Nichols et al. 2008). In that section, we argued that the only plausible explanation of the observed growth pattern is the signaling role of such peptides. Additional evidence comes from experiments showing an increased recovery of aquatic microorganisms in the presence of known signaling molecules (Bruns et al. 2002). While still hypothetical, the role of cell–cell communication in growth induction seems quite possible given the number of other microbial processes known to be regulated by quorum sensing (Waters and Bassler 2005; Camilli and Bassler 2006).

These considerations do not prove any of the hypothetical aspects of the scout model, but they do indicate that the assumptions involved are realistic, and that the proposed model is in line with observations from both environmental and molecular microbiology.

We now move on to explore several confusing and puzzling phenomena that this model can explain. We start with the Great Plate Count Anomaly. In this regard, we note that the three growth strategies proposed by the scout model (Fig. 7a–c) are not mutually exclusive, and while there may be species relying exclusively on one of them, there may be others that employ two or all three strategies in parallel. We think that if species in a community collectively employ these strategies in different mixes, the resulting behavior of the community, once moved to a Petri dish, will be similar to what is empirically observed in microbial cultivation: a small percentage of cells will form colonies, whereas the majority of them will not. The next section explains this in more detail.

6.2 Model Explains the Phenomenon of Uncultivability, and More

In Sects. 3–5 we described several observations that were difficult to reconcile with explanations for the Great Plate Count Anomaly offered to date. The scout model introduced here attempts to incorporate these observations. The general uncultivability of MSC33, for example, may exemplify all three proposed growth strategies. Suppose that MSC33 cells taken from the environment were dormant (there are several reasons why this could be so: dormancy could have been induced by the virtue of transferring the cells from in vivo to in vitro conditions, e.g., as a result of mixing them with warm agar or exposing them to the Petri dish environment; or else only (preexisting) dormant cells survived such mixing). According to the model, and in line with empirical observations, dormant cells of MSC33 would not grow unless provided with appropriate signaling compounds. When none are present, the only entity capable of growth is a scout cell, the presence of which is a chance event. By experimenting with MSC33 long enough, this chance event materialized, and we observed the first Petri dish reared colonies of MSC33, which we termed MSC33c. Used as a helper, the latter induced MSC33 to grow because it (MSC33c) produced growth-promoting signals. This growth induction could be replicated in the presence of other species (e.g., MSC105) because the latter, too, is a producer of a growth signal. In short, the model explains that some microorganisms, such as MSC33, have certain natural growth strategies, and once moved to the Petri dish they attempt to replicate these strategies as best as they can. In the case of MSC33, such replication results in its general uncultivability in vitro, interrupted by sporadic appearance of cultivable variants, capable of inducing growth of the parental population.

The model further explains that there may be two sources of domesticated variants in the otherwise largely uncultivable environmental populations.

The first source is the scouts, or cells that exited the state of dormancy due to stochastic variations in the mechanism controlling dormancy. This exit may be equally probable at any given point in time; whether or not a plated population forms colonies in vitro depends on the whether or not scout cells are present, which is in turn determined by the number of cells in the population and the rate at which they wake into scouts. Every dormant population that is abundant enough is expected to contain a certain number of cultivable scouts, whereas less abundant populations may lack scouts at some points, and “produce” them at others. Apart from the domestication phenomenon, this explains why, in a long-term cultivation experiment, we kept recovering the same species over and over again (Sect. 5). It also explains why a culture that took months to grow in a Petri dish, and thus gave every indication of being (very) slow growing, upon subculturing, no longer appeared “slow,” and instead grew like a weed. The appearance of a colony after a long period of incubation does not imply that these cells have an exceedingly long (and constant) generation time. In our view, it might just as well be a combined result of original dormancy, stochastic awakening at a random time point, and fast growth thereafter. We do not question the existence of genuinely slow growing species, but point out that these may represent only a fraction of strains that at first appear as such (i.e., those strains that initially required months to exhibit signs of growth).

Second, in populations exploiting the signaling scout strategy, it is reasonable to expect appearance and selection for overproducers of the signaling compound(s).

Such overproducers may be independent of the quorum-sensing-like regulation of growth of their parental populations, and will then appear in vitro as perfectly cultivable microorganisms.

It is possible that some of our domesticated strains (Sects. 3.1, 4) started from scouts of dormant populations, while others came from variants that overproduced signaling compounds. If this happened in our growth experiments, it must have happened in other cultivation studies as well. We therefore argue that some stocks in the existing type culture collections, such as ATCC, may be precisely such "overproducers," formed in the natural environment, and sampled by chance. If so, these cultivable strains are (genetic?) variants of the naturally occurring strain. One implication is that the wild-type cells, if sampled and plated, might appear uncultivable in vitro – even though cultivable strains are available from culture collections with identical 16S rRNA gene sequences.

We think that, from the point-of-view of selective advantage, the proposed growth strategies represent a rather clever microbial tool kit. Indeed, a system of randomly awakening scouts would provide a population with a simple mechanism of initiating growth in response to favorable conditions, without the need for detecting the actual moment environmental conditions shift to favorable ones. The scout simply grows, which ensures the reproduction of the genotype, or it dies. The latter is not necessarily a significant loss, even for populations of a small size. For example, if in a clonal population of just a thousand cells, on average one cell per day exits dormancy, and under adverse conditions survives for 10 days, then such a population (1) has 1% of its members continuously scouting the environment and checking if the conditions are appropriate for growth, (2) is capable of quick growth response to such conditions, and (3) can afford "scouting" for over 3 years before going extinct. In the end, this strategy enables this small population to safely reproduce even if there was just one day favoring growth during these 3 years! Arming the scout with a signaling function makes the strategy even "smarter," because then the dormant part of the population is able to massively respond to a favorable environmental event. Adding the signaling function to neighboring species has a potential of making this microbial growth strategy wiser still. Microorganisms have evolved exquisite mechanisms for joint use of available resources, some of which are well known (Boetius et al. 2000; Schink 2002). If a species senses the growth of its syntrophic partner and responds in kind, yet does not grow when such partners are absent, it would further coordinate the syntrophic activities and ensure that the metabolic potentials of the partners will be realized as meant by their coevolution, instead of staying unemployed in separation.

The model is not without shortcomings. Apart from the hypothetical nature of the postulated scouts and dormancy state, it appears to have a certain difficulty explaining a paradox mentioned in Sect. 4. Specifically, if the dormant (uncultivable) populations contain active (cultivable) scouts, why do we not see those in Petri dishes? And if we do, why have microbiologists not cultivated the bulk of natural microbial diversity yet?

We believe that this difficulty is only apparent, and in fact, the model is in accord with microbiological practice, for the following reason.

In a typical cultivation study, a Petri dish is inoculated with about 10^4 cells, which ordinarily results in several dozen to a hundred colonies. These usually belong to at most a few dozen species, and often fewer. Some of these species are represented multiple times, whereas the majority are typically rare. Does the scout model account for this?

We do not know the species' frequency in nature, but if the rRNA gene surveys are any guide, then a typical inoculum likely contains hundreds of variably represented species. The expectation is that an overwhelming majority of species among the inoculum's 10^4 cells are remarkably rare (Pedros-Alio 2006; Sogin et al. 2006; Huber et al. 2007), down to a single cell per species. Low abundance suggests inactivity, and so this (largest) part of the inoculum's species richness is a good candidate for being dormant. According to the first two strategies proposed (Fig. 7a,b), these will grow in vitro if stochastically awoken into scout cells. If the probability of having a scout is 0.1–1%, then only one out of 100–1,000 species, represented by a single cell each, will grow in a Petri dish. Therefore, a single Petri dish would be expected to contain fewer than a few dozen colonies of the rare species. Some of those will likely remain undetected (e.g., they may not grow before the dish is overtaken by the weed species, or their colonies are small and easy to miss, etc.). If the inoculum contained species relying on the third strategy, the number of isolated species will be even lower since these are not expected to grow unless other specific species grow first. All things considered, the model suggests that, in a standard cultivation study, one should expect to detect and isolate about 10–30 rare species, and a few more abundant species, per single Petri dish. Under the scout model, the remaining richness of the inoculum will appear uncultivable. This scenario is not unlike what is actually observed in real cultivation studies. Therefore, in this interpretation, and with this hypothetical percent of scouts, the model is in line with empirical observations.

These considerations lead us to the following hypothesis: a significant portion of colonies grown in a cultivation study from an environmental inoculum are in fact progeny of scouts, and a typical rate of recovery (0.1–1% of cells and species inoculated) is close to the percent of scouts in a dormant population, averaged over all populations present. Looking at microbial isolates on a Petri dish from this angle explains why replicate dishes typically have somewhat different composition, and why some species that grew well in one dish may be very difficult to reisolate even when the same method is applied to the same environment. It also explains why every now and then there are reports of significantly novel microorganisms cultivated using "just" the standard approaches, implying that in the past these approaches failed to isolate the microorganisms in question. The apparent randomness of microbial discovery may thus be a result of the rarity of some groups in nature, and the stochastic appearance of scout cells in their populations. We note that by itself the rarity of some groups in nature may not suffice as an explanation for low in vitro recovery: with (many) hundreds of rare species in each inoculum, the rarity alone fails to account for the fact that about 99% of species did not grow.

To summarize, we tend to explain a typically low colony and species recovery in standard media as a game of numbers: (1) the rarity of some microbial groups in environments sampled for cultivation so far, which is (2) magnified by infrequency of the scouts; (3) predominance of weeds, either prohibiting or masking the growth of scouts; and (4) reliance of some species on growth signals.

These explanations may have interesting implications for microbiology and health sciences, discussed in the next section.

6.3 The Implications of the Scout Model

6.3.1 What can be Domesticated, and How

In Sects. 3 we presented evidence that a large portion of uncultivable species can be grown in vitro because these species spontaneously domesticate and/or they can be grown in coculture with helper strains. In Sects. 6.1 and 6.2 we explained these observations by proposing the scout and signaling growth strategies. An implication of the model is that natural populations of microorganisms should be expected to contain, with certain probability, scout cells and cells overproducing the otherwise essential signaling compounds. Both types of cells are cultivable in vitro, and so it follows that, in principle, it should be possible to grow a significant number of novel species only using fairly trivial media. However, as we argue in Sect. 6.2, calculations based on the likely frequency distribution of species in environmental samples suggest that the actual odds of getting a significant portion of a sample's richness through cultivation are not high. Can these odds be improved?

In several published and unpublished studies, we employed a method of growing environmental samples in a series of sequential in situ incubations inside diffusion chambers (or ichips; Sect. 2). Once the grown material was inoculated into standard Petri dishes, we observed a score of novel species that were not detected by using more traditional methods. We see this as one example of changing the odds, due to a cumulative effect of several favorable factors. First, species sampled at the time they were actually active in situ, might continue their growth in diffusion chambers, leading to an increase in their biomass. Even if moving them into a Petri dish was stressful, and killed nondormant cells, or else induced dormancy in active cells, their larger populations would still produce larger number of scouts, improving the chances of growing and detecting at least some colonies in vitro. Second, those species that depend on growth-signaling compounds, whether originating from their kin or neighboring species, might grow in the chambers because diffusion provided those compounds during in situ incubation. Several passages through a number of diffusion chambers would further increase their biomass, thus increasing the probability of raising a (genetic?) variant independent of the signal. This would again result in obtaining cultures that might be difficult to isolate by standard methods. Third, once placed into the diffusion chambers and incubated in situ, some of the species that were dormant in the sample might find themselves in the vicinity of a new microenvironment that is conducive of growth. These, too, would exhibit a better probability of forming a colony in a Petri dish because, by virtue of increased cell count, they would produce either more scouts or more signal-independent cells, or both.

Perhaps there may be other reasons why, once the diffusion chamber-reared cells are inoculated into traditional media, the subcultured species seem different vis-à-vis what is obtained by standard approaches. Irrespective of what constitutes uncultivability, the empirical observation remains that one to several rounds of sample incubation in the diffusion chamber helps recovering novel species. Therefore, the method of sequential incubation in diffusion chambers or ichips, followed by inoculating their contents into standard Petri dishes, can serve as one method for accessing (domesticated) microbial novelty.

6.3.2 Scout Model Reinterprets Several Basic Microbiological Concepts

The scout model appears relevant to other microbiological concepts and phenomena. This section will focus on four: microbial growth curve, persister cells, latent infection, and viable but nonculturable cells.

Microbial growth curve. When a microbial culture grows in a closed system, it goes through several phases. Buchanan (1918) provided what is likely the earliest formal description of such phases, of which he recognized seven. Delpy et al. (1956) distinguished the currently more familiar log-, exponential, stationary, and death phases. Curiously, the earliest interpretation of the last phase was that total and viable counts are often inaccurate (Wilson 1922). The more recent interpretation of death phase is a period when the death rate exceeds the growth rate. This interpretation creates a rather dreary image of microbial culture in the late stationary phase, because of the perceived lack of activity, and predominance of death.

Pioneering works by Kolter's research group changed that perception (Zambrano et al. 1993). These works firmly established that at least some cells are actively multiplying and evolving during the late stationary phase. This leads to the selection of GASP (growth advantage in stationary phase, Zambrano and Kolter 1996) phenotypes whose fitness exceeds that of the parental population. Longer-term experiments showed a continuous evolution of GASP phenotypes, with the progressively increasing fitness, for a period of at least several weeks (Finkel and Kolter 1999).

This changed the view of late stationary and death phases as dynamic and full of microbial activities. Curiously though, as evidenced by a few available long-term experiments, after weeks of decline in the viable cell count and evolution of GASP phenotypes, this count actually appears to stabilize, and then stay essentially unchanged for years (Finkel et al. 2000). Note that at this point the initial load of nutrients is likely to have been exhausted. Why do then these cells not die out, and instead remain viable for months? Such persistence appears somewhat paradoxical. Maintaining viability, and capacity to quickly respond to fresh media, implies activity, and thus a source of energy and nutrients, the source supposedly absent at the time. Dead cells, and nutrients leaking from them, being a good resource for cannibalism at the time of decreasing cell count (Postgate and Hunter 1962), are unlikely to play the same role when there is no appreciable decline in the number of viable cells.

In the end, microbiologists face an empirical observation of a small number of cells that remain viable under conditions that do not seem to permit such persistence.

We note that such survival appears miraculous only if we habitually view the situation as static, that is, if we assume it is *the same* cells that remain viable for months. The paradox quickly disappears under the scout model, which postulates two things. First, starting from the onset of unfavorable conditions at the beginning of the stationary phase, environmental challenges induce some cells into dormancy. Importantly, the pool of dormant cells remains "invisible" as dormant cells do not grow, and mere plating cannot distinguish between dead and dormant cells. Second, the dormant cells stochastically produce scouts. During the first weeks following the stationary phase, the scouts may be difficult to detect as they are outnumbered by the cells with GASP phenotypes, but they become more prominent as the viable count drops further. The scout model predicts that, when the viable cell count drops to its minimum, and remains at this level for an extended period of time, this viable count is the count of scouts.

Scouts are active cells. They will not maintain their activity for long under severe starvation. During their lifespan, however, they remain viable and will form colonies if plated. The balance between the rate of the scouts appearance and disappearance, due to their random reawakening and death, then determines how many active, viable cells would be present in the culture at any given time. If the rate of scout formation is low (e.g., 0.1–1% per day; see our hypothesis in Sect. 6.2), it will take months to significantly impact the number of dormant cells, and by extension the absolute number of scouts. Therefore, for months there will be little decline in the count of viable cells in the culture, in accordance with the empirical observations. The model does not ascribe any inexplicable properties to the cells miraculously surviving during these months, but rather reassigns the function of survival to invisible dormant cells. Dormancy explains how population survives in the absence of energy and nutrients; spontaneously formed scouts explain why the viable count is not zero. The scout model thus offers a dynamic interpretation of the microbial growth curve at its later stages, when only a few viable cells remain. Like the GASP phenomenon, that changed how both the stationary phase and the first weeks following it are perceived, the scout model portrays the later stages as a dynamic state of a balance between stochastic awakening of scouts, and their death.

Persister cells. The scout model offers a new perspective on the nature of persister cells, defined as a small percentage of cells that remain viable after exposure to a high dose of antibiotic (see Lewis, "Persisters, Biofilms, and the Problem of Cultivability" in this volume for a detailed treatment of persistence). The previous theories proposed that persister cells survive the antibiotic challenge because they are specialized dormant cells that preexist in microbial populations. After removal of the antibiotic challenge, the persisters grow and form visible colonies (indeed, this is how they were initially discovered and how they are detected and counted today). We note that this definition appears to be somewhat contradictory. Indeed, the initial designation of persisters as dormant cells preexisting in the growing culture implied that they did not grow in this culture, that is, did not respond to fresh nutrients. Yet after the antibiotic challenge the very same cells supposedly do respond to the same medium, and grow. It is not clear how a persister cell can be dormant enough to *not* grow in the presence of nutrients, remain dormant during the antibiotic challenge (and thus survive it), and then, once the challenge has been removed, suddenly become active enough to form colonies. The scout model resolves this apparent paradox, with important implications. The model suggests the following hypothetical scenario.

A growing culture does indeed contain dormant cells that originated in the previous growth cycle, but they may be much more numerous than the current theories suggest. These dormant cells would continuously produce scouts, randomly appearing before, during, as well as after the antibiotic challenge. Scouts formed before the challenge would not be detectable because they are outnumbered by other cells growing at the time. However, they are easily detectable after the antibiotic has killed these growing cells. The model proposes that what today are called "persisters" are in fact scouts, which are no more than a manifestation of a much larger population of dormant cells, all of which survived the antibiotic challenge. Because of that, it seems more appropriate to apply the term persisters to the dormant cells, not the scouts they produce. The dormant cells would not grow in the Petri dish, which makes them indistinguishable from dead cells, and this is why they might have been missed by researchers. What the researchers empirically observed in the Petri dish is but a small fraction of this population, i.e., the scouts. The size of this fraction is the result of dynamic balance between the rate of scout formation, and their death.

In summary, the scout model reassigns the role of persisters to the cells that are nongrowing, not detectable by traditional methods, and typically ignored in scientific studies. This is not an argument of mere semantics, for it has important ramifications for current models and research strategies. For example, if what are today called persisters are indeed scouts, then they are typical active cells, and as such may not be particularly informative of the origin of antimicrobial tolerance, as is now supposed, because they (1) represent only a fraction of the real persisters, and (2) have actually existed in the very state that provided for their survival. Instead, the key to understanding the nature of the phenomenon of persistence may be in the processes of initiation and maintenance of dormancy, and in the stochastic exit from this state.

VBNC cells. The concept of the VBNC state was introduced in Sect. 6.1 (see also Colwell, "Viable but Not Cultivable Bacteria" in this volume). Apparently synonymous to dormancy, the VBNC state remains a hypothesis. It was repeatedly challenged by others (Bogosian et al. 1998; Bogosian and Bourneuf 2001; Nystrom 2003), who suggested that VBNC cells did not exist, and were dead – or injured beyond repair – cells. The main rationale for this alternative interpretation was that the proponents of the VBNC idea were not able to unequivocally show the process of VBNC revival. Hard evidence for the latter proved difficult to produce. One challenge appeared to be in that all populations, claimed to be in the VBNC state, contained a small but relentless fraction of cultivable cells. The common view is that these cells confuse the picture by making it difficult to differentiate between the revival of VBNC cells, and a trivial regrowth of a few remaining viable cells. The scout model offers a different perspective on the process of revival and the challenges in detecting it. The model views VBNC cells as the equivalent of the model's dormant cells. If so, then the cultivable fraction of cells, which the VBNC researchers found so troublesome, are scouts. Rather than confusing the picture, this fraction might have been, somewhat ironically, the very evidence of the revival these researchers tried to prove. Is the revival of VBNC cells possible beyond the scouts they produce? The signaling scout model (Fig. 7b) suggests a positive answer: the majority of VBNC could be revived by exposing them to helpers, such as, for example, a growing population of the kin. Literature contains evidence that such induction may be possible in at least one of the model organisms in VBNC research, *Salmonella* (Santo Domingo et al. 2000). Our successful attempts to induce growth of MSC33 by application of the spent medium of MSC33c and other unrelated species (Sects 3.2, 4; Nichols et al. 2008) provide further experimental support for this idea.

Latent infections. Consider a small population of dormant cells of an infectious agent that escaped both the antibiotic treatment and the host's immune system. In Sect. 6.2, we argued that a tiny population of a thousand dormant cells can revive, via scouts, at any time during several years of dormancy. In the case of a pathogen, such revival means transformation of a silent infection into an active disease. Under the scout model, the possibility of such recurrence is determined by a balance between the formation of the scouts and the ability of the immune system to eliminate their progeny. When the pathogen succeeds, the relapse appears to the observer as a random event. Note a conspicuous similarity of this to latent tuberculosis. Interestingly, in the mouse model, reactivation of persistent mycobacteria has been characterized as "spontaneous" (Stewart et al. 2003). The scout model predicts precisely this spontaneous reactivation for any pathogen that is reasonably well shielded from the host's immune system. While the model cannot predict the specific individual at risk, it identifies possible weaknesses of the pathogen. If the molecular mechanism of scout formation were known, one could envision devising a method of lowering the in vivo frequency of such formation, possibly to the point of rendering the dormant population harmless. Alternatively, one could eliminate the bulk of dormant cells by artificially wakening all dormant cells into scouts, followed by treatment with the available antimycobacterial drugs, which today mainly target actively growing cells. Either way, if the scout strategy *is* an important element of microbial survival strategy, its molecular underpinning would appear to be the mechanism to resolve, leading to better tools for disease control.

7 Conclusions

This work focused on the Great Plate Count Anomaly and attempts to explain it. On the basis of original and literature data, the chapter introduced a new scout model of microbial behavior. The behavior is described by periods of activity during favorable conditions, periods of dormancy during environmental challenges, and the transition between the two that occurs in at least three ways. First is the scout scenario, which postulates the random awakening of dormant cells for the purpose of exploring the environment. If the scout finds the conditions conducive for growth, it begins dividing and starts a new population. Second, this new population may produce signaling compounds inducing the remaining dormant cells to grow. Third, the same induction may be achieved, via cell–cell signaling, through activities of other, cooccurring (and unrelated) species. We argued that if at least some microbial species exploited these three strategies in nature, then their growth on standard media would produce an impression of microbial uncultivability, and thus the Great Plate Count Anomaly. This model allows us to answer the question with which we began this work: what does it take to cultivate the "missing" 99% of microbial diversity? Will a simple multiplication of previous cultivation efforts suffice, or is a significant change in the cultivation efforts in order? Our conclusion is that the answer is not a simple "either" – "or" dichotomy. If the scout/signaling strategies are characteristic of environmental species, we should expect all such microorganisms to produce – sooner or later – domesticated variants capable of growing in vitro. However, this does not necessarily mean that "rolling up our sleeves" is a panacea for bringing all these species into culture. Domesticated variants of uncultivable species may be so rare in the standard Petri dish, and/or they may be so well masked by other species, that the effort required to cultivate the majority of environmental species by finding their scouts may make the enterprise unrealistic. To beat the odds, new alternative cultivation approaches are required that have biases different from the standard approaches. One such alternative is an in situ incubation of environmental species inside diffusion chambers, because this seems to increase the probability of finding domesticated cells of species uncultivable by other methods.

The model's importance may be in connecting several key microbiological "dots." Apart from the Great Plate Count Anomaly, the model reinterprets the microbial growth curve and events past the late stationary phase, the nature of "persisters," and explains latent infections. It suggests that these seemingly unrelated concepts and phenomena could be viewed as different manifestations of a single simple cycle: transition from activity to dormancy in response to unfavorable factors, and back via stochastic awakening into scouts.

References

Aoi Y, Kinoshita T, Hata T, Ohta H, Obokata H, Tsuneda S (2009) Hollow fiber membrane chamber as a device for in situ environmental cultivation. Appl Environ Microbiol

Avery SV (2006) Microbial cell individuality and the underlying sources of heterogeneity. Nat Rev Microbiol 4:577–587

Becskei A, Seraphin B, Serrano L (2001) Positive feedback in eukaryotic gene networks: cell differentiation by graded to binary response conversion. EMBO J 20:2528–2535

Boetius A et al (2000) A marine microbial consortium apparently mediating anaerobic oxidation of methane. Nature 407:623–626

Bogosian G, Bourneuf EV (2001) A matter of bacterial life and death. EMBO Rep 2:770–774

Bogosian G, Morris PJ, O'Neil JP (1998) A mixed culture recovery method indicates that enteric bacteria do not enter the viable but nonculturable state. Appl Environ Microbiol 64:1736–1742

Bollmann A, Lewis K, Epstein SS (2007) Incubation of environmental samples in a diffusion chamber increases the diversity of recovered isolates. Appl Environ Microbiol 73:6386–6390
Bollmann A, Palumbo T, Lewis K, Epstein SS (2009) Isolation and characterization of novel bacteria from contaminated subsurface sediments. ISME J
Bruns A, Cypionka H, Overmann J (2002) Cyclic AMP and acyl homoserine lactones increase the cultivation efficiency of heterotrophic bacteria from the central Baltic Sea. Appl Environ Microbiol 68:3978–3987
Buchanan RE (1918) Life phases in a bacterial culture. J Infect Dis 23:109–125
Buerger S, Hong S-H, Lucey K, Epstein SS (2008) Single-cell approach to microbial cultivation reveals an unusual growth strategy. In: Abstracts of the 11th International Society for Microbial Ecology Symposium, Cairns, Australia, August 2008
Butkevich NV, Butkevich VS (1936) Multiplication of sea bacteria depending on the composition of the medium and on temperature. Microbiology 5:322–343
Button DK, Schut F, Quang P, Martin R, Robertson BR (1993) Viability and isolation of marine bacteria by dilution culture: theory, procedures, and initial results. Appl Environ Microbiol 59:881–891
Camilli A, Bassler BL (2006) Bacterial small-molecule signaling pathways. Science 311:1113–1116
Choi PJ, Cai L, Frieda K, Xie XS (2008) A stochastic single-molecule event triggers phenotype switching of a bacterial cell. Science 322:442–446
Connon SA, Giovannoni SJ (2002) High-throughput methods for culturing microorganisms in very-low-nutrient media yield diverse new marine isolates. Appl Environ Microbiol 68:3878–3885
Davis KE, Joseph SJ, Janssen PH (2005) Effects of growth medium, inoculum size, and incubation time on culturability and isolation of soil bacteria. Appl Environ Microbiol 71:826–834
Delpy LP, Beranger G, Kaweh M (1956) Method of counting living bacteria. Ann Inst Pasteur 91:112–114
Dubnau D, Losick R (2006) Bistability in bacteria. Mol Microbiol 61:564–572
Epstein SS (2009) Microbial awakenings. Nature 457:1083
Ferrari BC, Binnerup SJ, Gillings M (2005) Microcolony cultivation on a soil substrate membrane system selects for previously uncultured soil bacteria. Appl Environ Microbiol 71:8714–8720
Ferrell JE Jr (2002) Self-perpetuating states in signal transduction: positive feedback, double-negative feedback and bistability. Curr Opin Cell Biol 14:140–148
Finkel SE, Kolter R (1999) Evolution of microbial diversity during prolonged starvation. Proc Natl Acad Sci U S A 96:4023–4027
Finkel SE, Zinser E, Kolter R (2000) Long-term survival and evolution in stationary phase. In: Storz G, Hengge-Aronis R (eds) Bacterial Stress Responses. ASM, Washington, DC, pp 231–238
Fuqua WC, Winans SC, Greenberg EP (1994) Quorum sensing in bacteria: the LuxR-LuxI family of cell density-responsive transcriptional regulators. J Bacteriol 176:269–275
Gardner TS, Cantor CR, Collins JJ (2000) Construction of a genetic toggle switch in *Escherichia coli*. Nature 403:339–342
Gray TRG (1976) Survival of vegetative microbes in soil. Symp Soc Gen Microbiol 26:327–364
Huber JA et al (2007) Microbial population structures in the deep marine biosphere. Science 318:97–100
Jannasch HW, Jones GE (1959) Bacterial populations in seawater as determined by different methods of enumeration. Limnol Oceanogr 4:128–139
Kaeberlein T, Lewis K, Epstein SS (2002) Isolating uncultivable microorganisms in pure culture in a simulated natural environment. Science 296:1127–1129
Kuznetsov SI, Dubinina GA, Lapteva NA (1979) Biology of oligotrophic bacteria. Annu Rev Microbiol 33:377–387
Maamar H, Dubnau D (2005) Bistability in the *Bacillus subtilis* K-state (competence) system requires a positive feedback loop. Mol Microbiol 56:615–624

McInerney MJ et al (2008) Physiology, ecology, phylogeny, and genomics of microorganisms capable of syntrophic metabolism. Ann N Y Acad Sci 1125:58–72
Nichols D et al (2008) Short peptide induces an uncultivable microorganism to grow in vitro. Appl Environ Microbiol 74:4889–4897
Novitsky JA, Morita RY (1976) Morphological characterization of small cells resulting from nutrient starvation of a psychrophilic marine vibrio. Appl Environ Microbiol 32:617–622
Novitsky JA, Morita RY (1978) Possible strategy for the survival of marine bacteria under starvation conditions. Mar Biol 48:289–295
Nystrom T (2003) Nonculturable bacteria: programmed survival forms or cells at death's door? Bioessays 25:204–211
Pedros-Alio C (2006) Marine microbial diversity: can it be determined? Trends Microbiol 14:257–263
Postgate JR (1976) Death in macrobes and microbes. In: Gray TRG, Postgate JR (eds) The survival of vegetative microbes. Cambridge University Press, Cambridge, pp 1–19
Postgate JR, Hunter JR (1962) The survival of starved bacteria. J Gen Microbiol 29:233–263
Rappe MS, Giovannoni SJ (2003) The uncultured microbial majority. Annu Rev Microbiol 57:369–394
Rappe MS, Connon SA, Vergin KL, Giovannoni SJ (2002) Cultivation of the ubiquitous SAR11 marine bacterioplankton clade. Nature 418:630–633
Santo Domingo JW, Harmon S, Bennett J (2000) Survival of Salmonella species in river water. Curr Microbiol 40:409–417
Schink B (2002) Synergistic interactions in the microbial world. Antonie Van Leeuwenhoek 81:257–261
Schloss PD, Handelsman J (2004) Status of the microbial census. Microbiol Mol Biol Rev 68:686–691
Sogin ML et al. (2006) Microbial diversity in the deep sea and the underexplored "rare biosphere". Proc Natl Acad Sci U S A 103:12115–12120
Staley JT, Konopka A (1985) Measurement of in situ activities of nonphotosynthetic microorganisms in aquatic and terrestrial habitats. Annu Rev Microbiol 39:321–346
Stevenson LH (1978) A case for bacterial dormancy in aquatic systems. Microb Ecol 4:127–133
Stewart GR, Robertson BD, Young DB (2003) Tuberculosis: a problem with persistence. Nat Rev Microbiol 1:97–105
Torrella F, Morita RY (1981) Microcultural study of bacterial size changes and microcolony and ultramicrocolony formation by heterotrophic bacteria in seawater. Appl Environ Microbiol 41:518–527
Waters CM, Bassler BL (2005) Quorum sensing: cell-to-cell communication in bacteria. Annu Rev Cell Dev Biol 21:319–346
Wilson GS (1922) The proportion of viable bacteria in young cultures with especial reference to the technique employed in counting. J Bacteriol 7:405–446
Winogradsky S (1924) Sur la microflore autochtone de la terre arable. C R Acad Sci 178:1236–1239
Xu H-S, Roberts N, Singleton FL, Attwell RW, Grimes DJ, Colwell RR (1982) Survival and viability of non-culturable *Escherichia coli* and *Vibrio cholerae* in the estuarine and marine environment. Microb Ecol 8:313–323
Zambrano MM, Kolter R (1996) GASPing for life in stationary phase. Cell 86:181–184
Zambrano MM, Siegele DA, Almiron M, Tormo A, Kolter R (1993) Microbial competition: *Escherichia coli* mutants that take over stationary phase cultures. Science 259:1757–1760
Zieg J, Silverman M, Hilmen M, Simon M (1977) Recombinational switch for gene expression. Science 196:170–172

Metagenomics and Antibiotic Discovery from Uncultivated Bacteria

Vivian Miao (✉) and Julian Davies

Contents

Abstract Metagenomic approaches are transforming studies of diversity and relationships among microbial communities from any source. Catalogs of their genes and gene clusters can be assembled and compared to provide phylogenetic and biochemical information. Here we review the progress of metagenomic approaches for the production of novel small molecules from environmental bacteria, be they cultivable or uncultivable, by means of the isolation and expression of clusters of genes required for small molecule biosynthesis. In recent years, genetic and technical advances have permitted the design of numerous vectors and surrogate hosts that allow the heterologous expression of cloned sequences, and environmental DNA libraries of large fragments (that may encode entire biosynthetic pathways for

V. Miao and J. Davies
Department of Microbiology and Immunology, Life Sciences Centre, 2350 Health Sciences Mall, University of British Columbia, Vancouver BC V6T 1Z3, Canada
e-mail: vmiao@telus.net

Microbiol Monogr, doi:10.1007/7171_2008_8

bioactive small molecules) can be made. Metagenomic cloning is thus a promising approach for mining the extensive chemical diversity of different microbial populations, in particular by removing the past constraint of cultivation and allowing equal access to the large proportion of the community comprising yet uncultivated members. While the potential is great, there remain many technical barriers and only a limited number of novel compounds have been obtained. Nonetheless, it can be expected that the pharmaceutical promise of microbial products can and will be successfully exploited in coming years.

1 Secondary Metabolites and Antibiotics

1.1 Introduction to Secondary Metabolites

Microbes have an interesting chimeric characteristic. On one hand, they are the root cause of millennia of morbidity and mortality in human, animal, and plant populations; on the other hand, they generate the only successful treatments for most of these diseases. Almost all living organisms (including higher eukaryotes) produce biologically active small molecules. In the microbial world they are often produced in late log/stationary phase of growth; hence the name *secondary metabolites* (SMs) (Table 1). There is a metabolic transition during bacterial growth when the common primary metabolites (sugars, amino acids, nucleosides, etc.) are used as precursors of more complex, modified bioactive compounds. There are millions of such compounds produced in the microbial world; however, the main interest in SMs is that a number of them have proven to be powerful antibiotics or possess

Table 1 Some secondary metabolites, their producers, and their biosynthetic gene clusters

Compound	Organism	Gene cluster size (kb)	References
Gentamicin	*Micromonospora echinospora*	35	Unwin et al. (2004)
Streptomycin	*Streptomyces griseus*	40	Piepersberg (1994)
Daptomycin	*Streptomyces roseosporus*	55+	Miao et al. (2005)
Erythromycin	*Aeromicrobium erythraea*	55	Brikun et al. (2004)
Chloroeremomycin	*Amycolatopsis orientalis*	72	van Wageningen et al. (1998)
CDA	*Streptomyces coelicolor*	82	Hojati et al. (2002)
Bleomycin	*Streptomyces verticillus*	85	Du et al. (2000)
Ramoplanin	*Actinoplanes sp.*	88	Farnet et al. (2002)
Avermectin	*Streptomyces avermitilis*	90	Ikeda et al. (1999)
Rifamycin	*Amycolatopsis mediterranei*	90+	August et al. (1998)
Rapamycin	*Streptomyces hygroscopicus*	107	Schwecke et al. (1995)
Nystatin	*Streptomyces noursei*	123	Brautaset et al. (2000)

various other therapeutic activities. The antibiotic era began with the discoveries of Fleming (penicillin) and Waksman (streptomycin) and the enormous therapeutic success of these antibiotics led to the rapid expansion of the pharmaceutical industry capitalizing on the discovery of not only antibiotics but other novel bioactive natural products produced by microbes. In a scant half a century, these microbial products have changed the world. In the year 2004, the world market for antimicrobials was upwards of $30 billion (Demain 2000).

The origins and evolution of SMs are not known; suffice it to say that they are old in an evolutionary sense. The biosynthetic pathways for SMs can be quite complex, involving many different types of molecular transformations. As one example, the structure of the aminoglycoside antibiotic streptomycin possesses three separate sugar moieties all derived from D-glucose; it requires some 30 enzymic steps to achieve these conversions (Piepersberg 1994). The two most common classes of SMs are the nonribosomal peptides and the polyketides that are the products of large multienzyme complexes: nonribosomal peptide synthases (NRPS) and polyketide synthases (PKS) (Strohl 1997). In a single producing strain, these biochemical functions may be juxtaposed to synthesize natural hybrid molecules involving the connected activity of the different pathways. Host-encoded modifications of SMs are also found, with suites of enzymes necessary for structural changes (methylation, epimerization, glycosylation, etc.) that generate many variations of different bioactive molecules. The diversity of structure and function found in SMs will continue to be a rich source of pharmaceutically active molecules for the foreseeable future.

To date, a significant number of microbial genera and strains that produce therapeutically useful compounds have been developed for pharmaceutical use. It is widely appreciated that SMs offer a huge distribution of molecular "space" and bioactivity and have clearly evolved to be potent bioactive molecules, although it is not known what the activities are for! Added to this, the natural roles of SMs are not known, although it has been suggested that many may be important in the cell–cell signaling necessary to maintain stable microbial communities (Yim et al. 2006b).

1.2 Where Will New Antibiotics Come From?

There is a great need for continuing studies and supplies of new SMs that have the potential for new applications in various aspects of medicine. The growing threat of antibiotic-resistant pathogens raises the concern that, without antibiotics active against these potent infectious agents, the treatment of infectious diseases would revert to the preantibiotic era. Synthetic compounds (chemical combinatorial libraries) as sources cannot compare with the high levels of specific bioactivity found with natural products (Verdine 1996). A continuous supply of antibiotics to treat resistant pathogens, and new drugs to treat cancer, diabetes, cardiovascular, and increasing numbers of other human diseases is imperative, if current therapeutic practices are to be maintained. The directed semisynthetic approach has had some success but it is often difficult to modify the structures of complex secondary metabolites, except by arduous synthesis.

Thus it is necessary to expand the search for SMs from different sources. Therein lies the problem! Given that some 99% of all bacteria are uncultivated, the vast majority of the microbial producers of bioactive molecules are out of the reach of traditional isolation technology. In addition, the search for novel microbial drugs will become increasingly difficult since detection of active compounds may require new approaches that are more sensitive and not necessarily directed towards any specific functional assay. What are the possibilities of developing efficient methods for harvesting novel molecules from organisms that cannot be cultivated in the laboratory? By what means will it be possible to use molecular genetic approaches to introduce meaningful structural modifications that improve the biological activity or pharmacology of existing molecules and to produce new and better drugs? Natural chemical biology offers many opportunities for small molecule regulation and modulation of biochemical processes in cells and organisms.

1.3 Microbial Diversity and Metagenomics

The world of microbial diversity presents extreme logistical and technical challenges, and although the molecular diversity of microbes is a promising field for research, accessing or mining this world is in its infancy. It is difficult to estimate the SM potential of any given environment. Bacterial strains of many different genera have the capacity to produce bioactive small molecules. Of course, not all the producing organisms are found in the same environment; it has been known for some time that different soils vary in their content of SM producers. Estimates have been based on the detectable antibiotic activity of microbial isolates in the laboratory, but there are likely a very large number of bioactive SMs in the environment that are not detected, since effects are concentration dependent, and many SM producers would go undetected. Direct chemical methods of screening would be more instructive but it would be a brave natural product chemist who would attempt to identify all of the SMs that were made in a single soil sample. In a gram of soil, the number of bacteria is estimated to be 10^9–10^{10} (Sandaa et al. 1999; Torsvik et al. 1990) depending on the soil, and the number of different taxa varies enormously with up to 10^4 g^{-1} from pristine forest soils (Torsvik et al. 1998; Torsvik et al. 1990). An initial determination of 1.6×10^4 taxa per gram soil in an N-fertilized field (Sandaa et al. 1999), was increased by more than 100-fold when recalculated to include rare taxa (Gans et al. 2005). The actinomycetes are recognized as the most prolific SM producers (Watve et al. 2001), and culturable members of the group often number 10^6–10^7 cfu per gram of soil. The *Streptomyces hygroscopicus* group and its derivatives produce over 500 SMs in laboratory studies (J. Berdy, personal communication). Genome sequencing has revealed 20 known or proposed SM biosynthetic gene clusters in *S. coelicolor* (Bentley et al. 2002), and 30 clusters in

S. avermitilis (Ikeda et al. 2003); the clusters in the latter correspond to 6.6% of its genome. If a typical actinomycete had even 10 such gene clusters, this would suggest that the number of SM biosynthetic pathways in a gram of soil could be upwards of 10^7–10^8. This ignores the many other prolific SM producers among the actinobacteria, other Gram-positive bacteria such as the *Bacillacaea* (Stein 2005), and the Gram-negative environmental microbes including the *Pseudomonads* (Visca et al. 2007) and the *Myxococci* (Bode et al. 2003; Wenzel et al. 2005). Since many biosynthetic pathways are cryptic or not expressed under laboratory conditions, their small molecule products are unidentified.

It was Norman Pace and a succession of coworkers who, beginning in the late 1980s, pioneered the approach of analyzing bacterial communities (from soils, hot springs, and other sources) by isolating total genomic DNA from complex environmental populations ("eDNA") and cataloging the components, both culturable and nonculturable. This was the microbial revelation of the twentieth century, since it demonstrated that bacterial communities are enormously complex and diverse. Evolutionary trees became jungles and a new era of microbiology was born. This shot-gun cloning method of examining communities is now known as *metagenomics*, a term coined by Jo Handelsman of the University of Wisconsin in 1998, as "the application of modern genomic techniques to the study of microbial organisms directly in their natural environments, by-passing the need for isolation and laboratory cultivation of individual species". While the original and continuing application is the analysis of microbial populations from many different sources for their diversity of genera, species, subspecies, enzymes, and other characteristics, investigations of soil, marine, and intestinal populations have revealed many surprises in the past 10 years. The application of metagenomic approaches is the most likely approach to succeed in rescuing biosynthetic pathways from noncultivatable organisms in the environment.

It rapidly became apparent that metagenomic approaches could allow the isolation of the genes for novel enzymes from any environment (Daniel 2005; Langer et al. 2006; Lorenz and Eck 2005). It was then proposed that DNA fragments of a size large enough to encode the production of bioactive small molecules might be attainable (Fig. 1). More importantly, if these gene clusters could be expressed in a heterologous host, it would provide a direct route to the production of bioactive SMs such as antibiotics. Thus the entire SM biosynthetic capacity of any environment would be available for drug screening and discovery. The prospect was very attractive and several small biotechnology companies were formed with this goal in mind. Investigations by these groups as well as laboratories in academia over the last decade have provided useful "range-finding" data to probe how well metagenomics could be engaged as a tool in natural product drug discovery. Individual programs set their own mandates, and bioactivity was generously defined, but many ideas touching various aspects of the yet-to-be-delineated discovery process were broached during this exploratory phase. The collective efforts generated some promising proofs of concept, revealed knowledge and methodology gaps, and identified some critical areas to improve in this emerging technology.

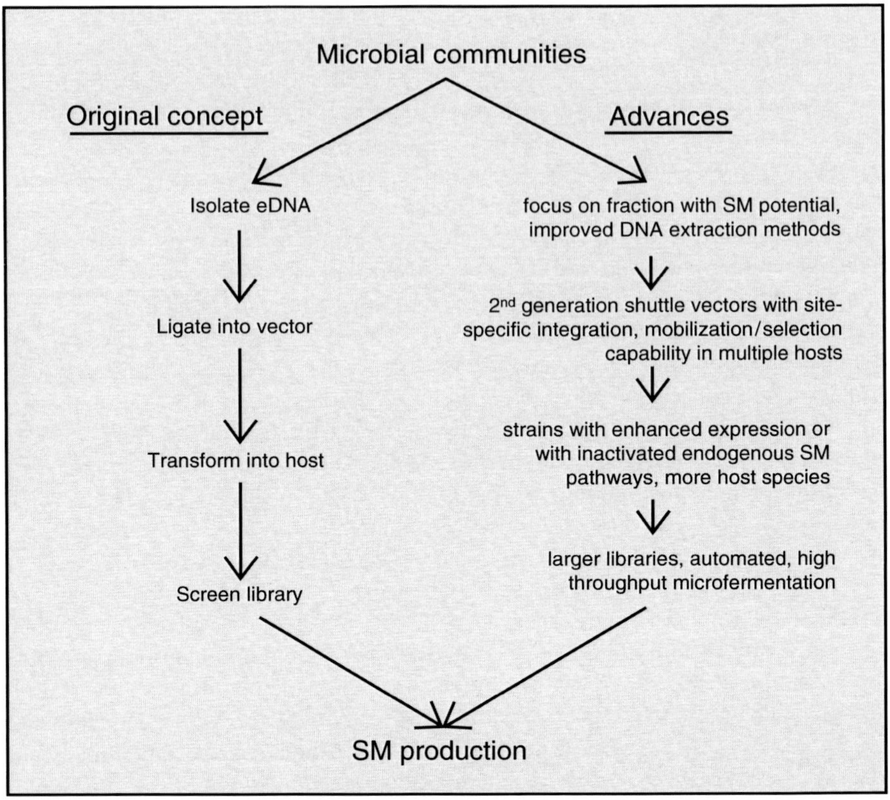

Fig. 1 Metagenomic approach to discovery of novel secondary metabolites (SMs). An environmental DNA-based method is not restricted by the ability to isolate and culture microbes, and allows access to the SM production capabilities of entire microbial communities. For retrieval of SM biosynthetic pathways, eDNA fragments large enough to contain intact polyketide synthase (PKS) and nonribosomal peptide synthase (NRPS) or other pathways are cloned into high-capacity vectors that can be introduced and maintained in tractable surrogate hosts to create libraries. Expression of the cloned pathways by the host leads to production of SMs that can be identified when the library is screened for antibiotic activity

2 Early Metagenomic Experiments with Soil DNA for Drug Discovery

2.1 *hsp70*, *gyrA*, Cellulases, and Beyond

The earliest work to follow on the lead provided by the 16S ribosomal gene studies of Pace and others was confirmation that numerous other useful and novel DNA sequences were present and could be similarly accessed from microbial communities in the soil. A phylogenetically diverse clone library of heat shock protein (*hsp70*) gene sequences was amplified from soil eDNA (Yap et al. 1996); a collection of fragments from unique cellulase genes was likewise obtained (Radomski et al. 1998). Sequences associated with antibiotic resistance or SM biosynthesis were also identified. New *gyrA* gene fragments that carried mutations associated with reduced sensitivity to fluoroquinolones were found in eDNA, raising the interesting possibility that resistance genes to even synthetic antimicrobials existed among microbial communities in the environment (Waters and Davies 1997). Waters et al. (1997) used degenerate primers targeting PKS genes to amplify novel sequences encoding catalytic domains of these important SM biosynthetic enzymes from eDNA. Other work showed that small, complete genes from a PKS biosynthetic gene cluster amplified from soil eDNA could be functionally integrated into a biosynthetic gene cluster to modify known secondary metabolites (Seow et al. 1997). Determination that the amplicons from eDNA in that study were all distinct, each tagging a hitherto unknown pathway, underscored the tremendous possibilities for genetic and chemical diversity, should the entire pathway be accessible.

2.2 eDNA Libraries and Heterologous Expression

Despite their utility in detecting various genes in eDNA, PCR-based sequence recovery platforms are limited in diversity by primer specificity, require other methods to obtain full-length genes, and remain constrained in the size of the DNA fragments that can be obtained. Libraries of eDNA that could be screened genetically, functionally, or by both means were recognized at the onset as essential for finding new biosynthetic pathways. Many studies investigated methods for isolating eDNA of a purity and size suitable for library construction (e.g., Miller et al. 1999; Robe et al. 2003; Zhou et al. 1996); some were concerned with the question of whether the recovered eDNA retained a proportional representation of the genotypes existing in the source environment (e.g., Courtois et al. 2001; Martin-Laurent et al. 2001). DNA was isolated from bacterial cells lysed in situ ("direct isolation") or from cells separated from soil ("indirect isolation") (Hopkins et al. 1991), cloned into general utility vectors, and arrayed or pooled. Functional screening was mainly done in the cloning host, *E. coli*, and activity was detected using indicator organism overlays (Rondon et al. 2000).

2.3 Signals, Resistance, Turbomycins, and Terragines

Long-chain *N*-acyl amino acids (Brady and Clardy 2000, 2005b) that likely have a role as components of cell signaling and quorum-sensing agents in microbial community communications were among the first bioactive molecules recovered from expression of eDNA in *E. coli*. Although the compounds were previously known, their biosynthetic origin was not. The gene responsible, encoding an *N*-acyl synthase, was identified and sequenced; isolation of other representatives of this gene family from additional eDNA libraries then allowed assignment of function to an unknown open reading frame (ORF) annotated in the genome of a cultured bacterium (Brady et al. 2004). Similarly, a known isocyanide-substituted indole containing compound was isolated from eDNA by activity screening (Brady and Clardy 2005a); further analysis of the metagenomic clone led to the recognition of the first isocyanide synthase (Brady and Clardy 2005c). Other known pigmented compounds or close derivatives thereof with activity found by screening cloned eDNA against *B. subtilis* included violacein (Brady et al. 2001), produced by a four gene biosynthetic pathway similar to that from *Chromobacterium violaceum* (August et al. 2000); turbomycins A and B (Gillespie et al. 2002) as well as turbomycin-like compounds synthesized by eDNA encoded 4-hydroxyphenylpyruvate dioxygenase; and indirubrin (MacNeil et al. 2001).

Functional screens for antibiotic resistance were conducted on the *E. coli* eDNA libraries as well. These not only indicated a wide distribution of putative resistance genes in the environment but also, given the known linkage of resistance with biosynthesis genes in antibiotic producers (Seno and Baltz 1989), could indirectly lead to discovery of new pathways. Fragments of eDNA conferring resistance of *E. coli* to kanamycin or other aminoglycosides were identified at low frequencies in several eDNA libraries (Courtois et al. 2003; Riesenfeld et al. 2004a), but the genes involved, encoding aminoglycoside modifying or tetracycline efflux proteins, were unlinked to biosynthetic genes. In the one case studied (Courtois et al. 2003), the putative resistance gene did not function in *S. lividans.* This finding, and the low GC composition of the eDNA sequences, suggested that the resistance genes did not originate from aminoglycoside producing actinomycete strains.

Other studies, conducted using both *E. coli* and *S. lividans,* focused on demonstrations of heterologous expression of new chemical entities. The terragines (Wang et al. 2000), the first novel structures obtained by expression of a replicative eDNA cosmid library in *S. lividans*, were identified by HPLC analysis of the fermentation products of 1,000 recombinant strains. Another study with a smaller number of recombinant *S. lividans,* but employing a variety of fermentation media, reported two new fatty acid dienes (Courtois et al. 2003). These results demonstrated the possibility, as suggested previously by PCR studies, of using an actinomycete host for heterologous expression of eDNA.

3 Progress in Cloning SM Biosynthetic Pathways from eDNA

3.1 SM Pathway Focused Model

The organisms best known and studied for antibiotic production are the soil actinomycetes. The polyketide and nonribosomal peptide SMs produced by the actinomycetes have been the corner stone of natural product–based drug discovery. Production of these compounds is encoded by biosynthetic gene clusters that include not only the large genes encoding the PKS(s) or NRPS(s) enzymes but also a variety of pathway specific genes contributing modification, regulatory, and other functions. Recovery of this type of gene cluster would provide compelling proof of concept that metagenomics could be relied on as a platform for generation of drugs based on new microbially derived SMs. However, despite success with single genes and small gene clusters (Brady et al. 2001), and ample demonstration of novel NRPS and PKS gene fragments in the environment (Seow et al. 1997; Waters et al. 1997; Wawrik et al. 2005) as well as in some eDNA libraries (Courtois et al. 2003; Ginolhac et al. 2004), to date, no complex biosynthetic gene cluster has been recovered from eDNA.

One reason that eDNA libraries fell short of the target for drug discovery may be that they were often (intentionally) "general purpose" libraries, insofar as the vision was to embrace a largesse of diverse uncultivated genomes and exploit them toward a variety of purposes (Rondon et al. 2000; Schloss and Handelsman 2003; Streit and Schmitz 2004). One proposed advantage from a drug discovery perspective is that this offered a greater opportunity to reveal new chemical classes of SM not derived from typical types of biosynthetic pathways. To enable the generalist approach, however, source selection and processing, cloning criteria and expression system, screening platforms and throughput were all areas where compromises were accepted. The discovery of interesting bioactive SMs that may yet be developed for new therapies based on modulation of intercellular communication among bacteria has borne out this hope, but there is still an enormous gulf between the identification of a natural product that has activity, and one that can be progressed as a viable lead by the pharmaceutical industry. A more conservative but more immediately productive path may be to revise the original "all-inclusive" model, and partition the metagenome to focus on capturing relevant "SM pathway containing" fractions. In addition, a greater collaboration between technologists and scientists in other fields who have a complementary focus of interest would be beneficial. Metagenomics-based technology is continually evolving better strategies and tools for cloning and expressing genes, screening, tracking, analysis and identification, and taking in new information from other sources. For example, genome sequencing and SM pathway cloning and characterization studies have provided a much clearer idea of the abundance of these gene clusters in bacteria, and their size range (McAlpine et al. 2005; Zazopoulos et al. 2003). Thus, combining what is known about SM antibiotics and their sources with collective expertise in metagenomics might more successfully target complex SM biosynthetic pathways that can be heterologously expressed in a tractable surrogate host.

3.2 "Handicapping" the Metagenome

A metagenomic drug discovery platform expends considerable resources on a sample comprising very few grams of soil; that sample should therefore be selected with care! The choices should be based on an integration of all relevant knowledge, tapping into assessments of target taxa diversity, richness, and representation, and flagged by genetic screens for the presence of PKS, NRPS, or other expected amplicons; geographic as well as seasonal variations in SM producer populations in the communities should also be a consideration. Guidance from microbial ecologists and microbiologists with experience in drug discovery would confer a greater degree of sophistication in identifying quality sources (Horan 1994). Because bacteria are distributed in microhabitats, and spatial variability is an important consideration, identifying and pooling a few teaspoons of soil from the best soil horizons or rhizosphere volumes, for example, would be more efficacious than a shovelful of bulk soil, with respect to both overall efficiency of eDNA extraction, and resulting concentration of target eDNA. Pooling dilutes local rarities, but could be used in combination with a variety of compensatory "enrichment" steps (below); likewise, pooling from different sources, combined with enrichment processing, would increase the sampling space for obtaining the desired population. Once the sample is collected, every effort should be made to sift the microbial community for the taxa of interest. A variety of selective or nutrient media and conventional microbiology techniques (Kurtböke et al. 2003; Otoguro et al. 2001; Takahashi and Omura 2003) have been used to enrich various groups of bacteria in traditional culture-based searches for new antibiotic producers; these might be adapted to process soil samples prior to DNA extraction (Bertrand et al. 2005). For example, to increase the relative proportion of spore-forming actinomycetes at the expense of other genera, one might incubate the soil in a selective antibiotic cocktail detrimental to undesirable taxa, dry the soil to encourage spore formation, and finally extract spores preferentially from the biomass, leaving cells of antibiotic-sensitive, non-spore-forming taxa behind.

The perennial concerns during eDNA isolation include co-purification of humic substances that inhibit PCR or cloning enzymes, shearing of DNA, and fidelity of representation. The problem of humic substances, primarily an issue for early solution-phase isolation methods, is now minimized by a variety of means, including special gel formulations and electrophoretic systems (Pel et al. 2007; Quaiser et al. 2002). The many studies concerned with methods to purify DNA from soil indicate that direct isolation results in a higher recovery rate, but the DNA may include eukaryotic and extracellular DNA (Ogram et al. 1987; Steffan et al. 1988) and will be more degraded by shearing. In contrast, indirectly extracted DNA (Berry et al. 2003; Nalin et al. 2006), while more laborious in preparation and generally obtained in lower quantity, tends to be larger in size. The latter is therefore more suitable for cloning large gene clusters, until new methods are found for isolating high-molecular-weight DNA by direct extraction. The question of representational bias (Martin-Laurent et al. 2001) associated with the choice of isolation method is only relevant insofar as maintaining or enriching the fraction containing the target genomes with SM gene clusters. Preferential recovery of DNA from GC-rich taxa such as actinomycetes during cell lysis by using lysozyme (Treusch et al. 2004) or differential purification based on GC composition, e.g., with bisbenzimide, may be applied at later stages of DNA purification. The final purified eDNA need not be "chromosome-sized", but only large enough for partial digestion or end-polishing and cloning into a large capacity vector. A stringent criterion for a minimum insert size in libraries, however, should be followed. The general-purpose eDNA libraries to date carry inserts often much below 50 kb (but the occasional heartening giant inserts have been noted); small insert size, compared to the size of some known SM biosynthetic gene clusters (Table 1), can by itself account for the failure of metagenomics to recover an SM encoded by eDNA. Deliberately added controls, or "spikes", such as a known titer of one or more representative marker organisms or some known DNAs, allow assessment of various stages of eDNA isolation (Frostegard et al. 1999) and library preparation. If the control included a sequenced, screenable SM biosynthetic pathway, then the entire process, from DNA isolation to SM production in an expression host, could be systematically monitored and improved.

3.3 Better Vectors

Most of the early screening of metagenomic libraries employed *E. coli* for expression. The choice was understandable given the convenience of cloning in *E. coli* and the need to examine large numbers of clones, but the chances for success in the expression of large biosynthetic gene clusters cloned directly from eDNA are slim. While it has been estimated (Gabor et al. 2004) that 40% of the genes from eDNA (as represented by 32 sequenced genomes of diverse microorganisms) might function in *E. coli*, experience has shown that a good deal of ingenuity in genetic engineering and skill are required to express large SM biosynthetic pathways in *E. coli* effectively (Mutka et al. 2006; Pfeifer et al. 2002; Pfeifer et al. 2003). Shuttle vectors and new host strains have now been constructed to redress a lack that may have contributed to past reliance on *E. coli*. Use of a shuttle vector is a more efficient general strategy than genetic screening of a library cloned in a conventional vector (Ginolhac et al. 2004; Liles et al. 2003) followed by pathway reconstruction (Sosio et al., 2000), or adjustment of the vector backbone in order to mobilize interesting clones into another organism; the latter would be impractical if there were many hits. Commonly used cosmids (Courtois et al. 2003; Miao et al. 2006), P1 or bacterial artificial chromosome (PAC and BAC, respectively) vectors (Martinez et al. 2004; Miao et al. 2005; Sosio et al. 2000), were fitted with markers for selection in both *E. coli* and other hosts, an origin of transfer for conjugative mobilization of the library, and a site-specific integration system from a phage, e.g., ΦC31, or plasmid, e.g., pSAM2 (Kuhstoss et al. 1991), to enable the insertion of eDNA into the chromosome of the expression host for stable maintenance. Smaller biosynthetic gene clusters can be captured in cosmids or fosmids, but multihost BACs provide the best chance for cloning intact pathways and moving them into a number of backgrounds with different capabilities for heterologous expression.

3.4 Better Hosts

S. lividans is the most frequently used expression host after *E. coli*. *S. lividans* is closely related to *S. coelicolor*, a familiar model for studies of development and secondary metabolism, and for which a genome sequence is available (Bentley et al. 2002).

Clones from most *E. coli* strains can be efficiently introduced into *S. lividans* by protoplast transformation or high-throughput conjugation (Martinez et al. 2004). Spontaneous (Hu and Ochi 2001) or engineered (Martinez et al. 2005; Okamoto-Hosoya et al. 2003) mutants with increased capacity for SM production and strains in which endogenous pathways (Martinez et al. 2004; Ziermann and Betlach 1999) were inactivated have been made. The latter reduces possible competition for metabolic precursors as well as background activity, and may simplify product isolation downstream (Penn et al. 2006). Interestingly, eDNA that increases production of host secondary metabolites have been observed (Courtois et al. 2003) (Davies, unpublished), and an enhanced production mutant was constructed by addition of a regulatory gene, *rep*, that originated from an eDNA library (Martinez et al. 2005). A Gram-negative strain, *Pseudomonas putid*a, has also been improved for heterologous expression by introduction of a chromosomal ΦC31 *attB* site, thereby making it compatible with any eDNA library built in a ΦC31-based vector (Martinez et al. 2004).

The development of vectors and hosts should be coordinated with each other as well as with considerations for other stages of the discovery process. The task of purifying and characterizing novel SMs often requires improvement of yield from the producing strain: this involves fermentation optimization but may also include strain development. The latter typically takes place after a product is recognized to be of high value and the target parameters are well defined; however, in a metagenomic discovery platform, strain development should be a component at an early stage, as the investment could be easily rationalized for facilitating not only library building and hit detection but also downstream processing and product isolation. Experience with strains that have been developed for model pathways of various classes provide a starting point to build on. *E. coli* strains with potential to express some pathways could be implemented as the primary cloning host, thereby offering an earlier, more convenient screening opportunity, and a panel of specialized expression hosts such as strains of *S. lividans*, *Bacillus subtilis* (Doekel et al. 2002; Eppelmann et al. 2001), *P. putida* and others (Wenzel et al. 2005; Wenzel and Muller 2005) could complement and enhance this by providing additional screening and production opportunities. A multiplicity of optimized surrogate expression hosts further raises the possibility that host-specific biochemical changes may lead to derivatives of the compound specified by the cloned pathway.

3.5 Library Size

The number of clones needed to capture a complete metagenome in the proverbial gram of soil has been periodically re-estimated: an early suggestion that 10^6 BAC-sized clone (Handelsman et al. 1998) was revised to 10^{11} (Riesenfeld et al. 2004b) to include rare taxa. The latter is justified in view of the realization that microbial diversity in soils may be much greater than previously thought (Gans et al. 2005; Neufeld and Mohn 2005). From the narrower focus of drug discovery, the diversity component that is most relevant is that of SM biosynthetic pathways. The library size for this would be smaller, with the degree depending greatly on the success of upstream target eDNA enrichment schemes as well as on the effectiveness of downstream processes to winnow the clones. If greater taxonomic diversity means more clonable pathways, then the higher estimates of diversity against the same total cell count in a gram of soil could in effect dereplicate some of the input DNA and have a positive impact on the construction of eDNA libraries. The same corollary would apply to estimates of the number clones needed to find a previously undiscovered antibiotic biosynthetic gene cluster (Baltz 2007). The most recent eDNA libraries, described in gigabases of eDNA, are already larger (Williamson et al. 2005), in keeping with the need to move to a scale that is more commensurate with the goals.

3.6 Updated Screening

The typical process for identifying bioactive SMs consists of growing microbes in batch (flask) fermentations in multiple conditions and nutrients, followed by exposure of whole or clarified broth, or extracts thereof, to various activity screens, and then chemical analysis and dereplication of hits. When activity was determined using susceptible strains, essentially all cellular targets are evaluated at once and penetration and in vivo activity are confirmed. This was supplanted by high-throughput, target-based (in vitro) screening for a time, but results to date are less than hoped for, and interest in the merits of the whole cell methods has revived. In order to process the enormous number of recombinants from expression libraries, the return to whole-cell screening must be updated to include automated procedures. New protocols could take advantage of concepts, instrumentation, and methods devised for target-based screening to achieve the requisite level of throughput.

Encapsulation of bacteria in alginate beads has been used as a scaled-down method for growing both natural (Zengler et al. 2002) and recombinant strains for screening (Silva et al. 2001); screens might be designed to use fluorescence-activated cell sorting in selected circumstances (Short 2000; Short and Keller 2004; Yun and Ryu 2005). Microfermentation in beads, coupled with bioactivity assays against an engineered test strain carrying multiple resistance genes against common known antibiotics is already the basis of a screen with a throughput of millions of natural strains per year (Baker et al. 2007; Baltz 2006).

The full potential of metagenomic production of SM can only be realized with the means for broad screening for bioactivity. The relationship between structure and activity of SMs is poorly understood; e.g., an antibiotic structure may be modified but a novel antitumour or cardiovascular activity may result, and quorum-sensing type molecules were not conceived of as antibiotics, until they emerged from a screen for bacteriocidality (Brady and Clardy 2005). An interesting system, designated METREX, consists of an expression host carrying a GFP reporter sensitive to compounds that induce quorum sensing; expression of eDNA library clones encoding genes for synthesis of bioactive small molecules that induced the reporter could then be isolated by fluorescence (Williamson et al. 2005). Development of other systems with new selective, specific, and sensitive testers (Yim et al. 2006a) plus high-throughput platforms are needed to provide options that fully exploit the potential within metagenomic libraries.

A preliminary report (Coeffet-LeGal et al. 2001) of shuttle cosmid libraries containing bacteria enriched from forest soils incorporated several of the above ideas. Spores were preferentially extracted from dried soil and incubated briefly in a liquid culture favoring actinomycetes; DNA was extracted, determined to be enriched for high GC eDNA by denaturing gradient gel electrophoresis, and archived into cosmid libraries; random library clones were then confirmed to be GC rich by restriction digestion. DNA from the estimated 4×10^4 member library (1.6 Gb equivalent) was collected from *E. coli*, the cloning host, pooled, and, after sucrose gradient purification to ensure collection of the largest DNAs, introduced into *S. lividans* by transformation and integration at the ΦC31 site. Spores from recombinants were encapsulated in alginate beads for fermentation and 14 of 2.4×10^5 derivatives were active when tested against *S. aureus*. Key benefits of this approach were that strains could be recovered from the alginate for further studies, and that the expression library could be fermented under a variety of other nutrient and environmental conditions and tested against more than one organism. Provided that scale-up fermentations reproduce the activities observed initially, the output from a metagenomic expression library of eDNA can then be segued into an existing natural product discovery pipeline, integrating with all of the optimized processes therein.

4 Conclusion

Attempting to develop a new paradigm for SM drug discovery, while at the same time trying to comprehend the still-unfolding enormity of the metagenomic world and to grasp it with a common set of tools, was an ambitious undertaking, and the successful cloning and expressing of complex biosynthetic gene clusters from eDNA is far from routine. The body of work to date has principally been exploratory, but each investigation had added to the sophistication of understanding and built up the toolbox needed to undertake the challenge properly. The technical elements in isolation of high-molecular-weight eDNA and the cloning of large fragments into second generation multihost shuttle vectors are basically worked out, and some high-throughput methods, e.g., microfermentation, coupled with more sophisticated screening systems are also in place. Expression of eDNA will be a critical area to develop in the near future, and generation of tractable hosts with enhanced expression capability is a good beginning. The production of new compounds or new activities, some from a combination of host-encoded and library-encoded functions, from the early eDNA libraries indicate that the principles of gaining access to uncultivable organisms via metagenomics are sound (Clardy et al. 2006); however, it is necessary to devise more dedicated strategies and integrated workflows, in order to progress to the next level of the technology and to work with complex SM biosynthetic pathways. The search for validation of soil metagenomic cloning in the near future could well be the successful cloning of one of the simpler NRPS- or PKS-derived molecules.

eDNA containing whole or partial pathways can in addition be applied to the following aspects of natural product discovery:

(a) Modification of existing pathways
(b) Yield improvement
(c) Structure modification
(d) Hybrid (combinatorial) pathways
(e) Reconstruction of complex biochemical pathways
(f) Novel modifications of existing structures
(g) Identification of resistance mechanisms
(h) Generation of analog families around core molecules

These objectives are all achievable through the successful application of metagenomics to single organisms, to mixtures of organisms (natural or designed) in vivo, or by extracellular manipulation.

To sum up, metagenomics is a powerful tool that is already having an enormous impact on the cataloging and analysis of microbial populations and in the foreseeable future will likely provide an inventory of the microbial population of the biosphere and help elucidate the structures of microbial communities. Whether this technique will be applicable to the intriguing problem of characterizing their communication networks and the natural roles of many millions of SMs remains to be seen. With the recognition of even greater metabolic diversity among terrestrial and marine microbes (Cavicchioli et al. 2007; Venter et al. 2004; Ward and Bora 2006), achievement of a robust metagenomic approach to isolate the biosynthetic pathways of complex molecules in both cultivable and noncultivable microbes and to manipulate them in more "friendly" hosts will contribute a revolutionary and powerful technology to natural-product drug discovery and modification. The resulting pharmacopeia can be expected to revolutionize medicine in the same way that antibiotics did in the past half century.

References

August PR, Tang L, Yoon YJ, Ning S, Muller R, Yu TW, Taylor M, Hoffmann D, Kim CG, Zhang X, Hutchinson CR, Floss HG (1998) Biosynthesis of the ansamycin antibiotic rifamycin: deductions from the molecular analysis of the rif biosynthetic gene cluster of *Amycolatopsis mediterranei* S699. Chem Biol 5:69–79

August PR, Grossman TH, Minor C, Draper MP, MacNeil IA, Pemberton JM, Call KM, Holt D, Osburne MS (2000) Sequence analysis and functional characterization of the violacein biosynthetic pathway from *Chromobacterium violaceum*. J Mol Microbiol Biotechnol 2:513–519

Baker DD, Hartland C, Monahan C (2007) High-throughput screening of natural products using an engineered drug resistant strain to discover novel classes of antibiotics. In: Society for biomolecular sciences 13th annual conference and exhibition, Montreal, Canada

Baltz RH (2006) Marcel Faber Roundtable: is our antibiotic pipeline unproductive because of starvation, constipation or lack of inspiration? J Ind Microbiol Biotechnol 33:507–513

Baltz RH (2007) Antimicrobials from actinomycetes: back to the future. Microbe 2:125–131

Bentley SD, Chater KF, Cerdeno-Tarraga AM, Challis GL, Thomson NR, James KD, Harris DE, Quail MA, Kieser H, Harper D, Bateman A, Brown S, Chandra G, Chen CW, Collins M, Cronin A, Fraser A, Goble A, Hidalgo J, Hornsby T, Howarth S, Huang CH, Kieser T, Larke L, Murphy L, Oliver K, O'Neil S, Rabbinowitsch E, Rajandream MA, Rutherford K, Rutter S, Seeger K, Saunders D, Sharp S, Squares R, Squares S, Taylor K, Warren T, Wietzorrek A, Woodward J, Barrell BG, Parkhill J, Hopwood DA (2002) Complete genome sequence of the model actinomycete *Streptomyces coelicolor* A3(2). Nature 417:141–147

Berry AE, Chiocchini C, Selby T, Sosio M, Wellington EM (2003) Isolation of high molecular weight DNA from soil for cloning into BAC vectors. FEMS Microbiol Lett 223:15–20

Bertrand H, Poly F, Van VT, Lombard N, Nalin R, Vogel TM, Simonet P (2005) High molecular weight DNA recovery from soils prerequisite for biotechnological metagenomic library construction. J Microbiol Methods 62:1–11

Bode HB, Irschik H, Wenzel SC, Reichenbach H, Muller R, Hofle G (2003) The leupyrrins: a structurally unique family of secondary metabolites from the myxobacterium *Sorangium cellulosum*. J Nat Prod 66:1203–1206

Brady SF, Clardy J (2000) Long-chain *N*-acyl amino acid antibiotics isolated from heterologously expressed environmental DNA. J Am Chem Soc 122:12903–12904

Brady SF, Clardy J (2005a) Cloning and heterologous expression of isocyanide biosynthetic genes from environmental DNA. Angew Chem Int Ed Engl 44:7063–7065

Brady SF, Clardy J (2005b) *N*-acyl derivatives of arginine and tryptophan isolated from environmental DNA expressed in *Escherichia coli*. Org Lett 7:3613–3616

Brady SF, Clardy J (2005c) Systematic investigation of the *Escherichia coli* metabolome for the biosynthetic origin of an isocyanide carbon atom. Angew Chem Int Ed Engl 44:7045–7048

Brady SF, Chao CJ, Handelsman J, Clardy J (2001) Cloning and heterologous expression of a natural product biosynthetic gene cluster from eDNA. Org Lett 3:1981–1984

Brady SF, Chao CJ, Clardy J (2004) Long-chain *N*-acyltyrosine synthases from environmental DNA. Appl Environ Microbiol 70:6865–6870

Brautaset T, Sekurova ON, Sletta H, Ellingsen TE, StrLm AR, Valla S, Zotchev SB (2000) Biosynthesis of the polyene antifungal antibiotic nystatin in *Streptomyces noursei* ATCC 11455: analysis of the gene cluster and deduction of the biosynthetic pathway. Chem Biol 7:395–403

Brikun IA, Reeves AR, Cernota WH, Luu MB, Weber JM (2004) The erythromycin biosynthetic gene cluster of *Aeromicrobium erythreum*. J Ind Microbiol Biotechnol 31:335–344

Cavicchioli R, Demaere MZ, Thomas T (2007) Metagenomic studies reveal the critical and wide-ranging ecological importance of uncultivated archaea: the role of ammonia oxidizers. Bioessays 29:11–14

Clardy J, Fischbach MA, Walsh CT (2006) New antibiotics from bacterial natural products. Nat Biotechnol 24:1541–1550

Coeffet-LeGal MF, Charman K, Dugay F, Nichols M, She K, Silva CJ, Williamson N, Brian P, Miao V (2001) Construction and screening of recombinant clones from the environment (abstract). In: International symposium on the biology of actinomycetes, Vancouver, August 5–9, 2001

Courtois S, Frostegard A, Goransson P, Depret G, Jeannin P, Simonet P (2001) Quantification of bacterial subgroups in soil: comparison of DNA extracted directly from soil or from cells previously released by density gradient centrifugation. Environ Microbiol 3:431–439

Courtois S, Cappellano CM, Ball M, Francou FX, Normand P, Helynck G, Martinez A, Kolvek SJ, Hopke J, Osburne MS, August PR, Nalin R, Guerineau M, Jeannin P, Simonet P, Pernodet JL (2003) Recombinant environmental libraries provide access to microbial diversity for drug discovery from natural products. Appl Environ Microbiol 69:49–55

Daniel R (2005) The metagenomics of soil. Nat Rev Microbiol 3:470–478

Demain AL (2000) Small bugs, big business: the economic power of the microbe. Biotechnol Adv 18:499–514

Doekel S, Eppelmann K, Marahiel MA (2002) Heterologous expression of nonribosomal peptide synthetases in B. subtilis: construction of a bi-functional *B subtilis/E coli* shuttle vector system. FEMS Microbiol Lett 216:185–191

Du L, Sanchez C, Chen M, Edwards DJ, Shen B (2000) The biosynthetic gene cluster for the antitumor drug bleomycin from *Streptomyces verticillus* ATCC15003 supporting functional interactions between nonribosomal peptide synthetases and a polyketide synthase. Chem Biol 7:623–642

Eppelmann K, Doekel S, Marahiel MA (2001) Engineered biosynthesis of the peptide antibiotic bacitracin in the surrogate host *Bacillus subtilis*. J Biol Chem 276:34824–31

Farnet CM, Zazopoulos E, Staffa A (2002) Gene cluster for ramoplanin biosynthesis. WO 02/32255 A2, Canada Patent 20,020,164,747, April 18, 2002

Frostegard A, Courtois S, Ramisse V, Clerc S, Bernillon D, Le Gall F, Jeannin P, Nesme X, Simonet P (1999) Quantification of bias related to the extraction of DNA directly from soils. Appl Environ Microbiol 65:5409–5420

Gabor EM, Alkema WB, Janssen DB (2004) Quantifying the accessibility of the metagenome by random expression cloning techniques. Environ Microbiol 6:879–886

Gans J, Wolinsky M, Dunbar J (2005) Computational improvements reveal great bacterial diversity and high metal toxicity in soil. Science 309:1387–1390

Gillespie DE, Brady SF, Bettermann AD, Cianciotto NP, Liles MR, Rondon MR, Clardy J, Goodman RM, Handelsman J (2002) Isolation of antibiotics turbomycin a and B from a metagenomic library of soil microbial DNA. Appl Environ Microbiol 68:4301–4306

Ginolhac A, Jarrin C, Gillet B, Robe P, Pujic P, Tuphile K, Bertrand H, Vogel TM, Perriere G, Simonet P, Nalin R (2004) Phylogenetic analysis of polyketide synthase I domains from soil metagenomic libraries allows selection of promising clones. Appl Environ Microbiol 70:5522–5527

Handelsman J, Rondon MR, Brady SF, Clardy J, Goodman RM (1998) Molecular biological access to the chemistry of unknown soil microbes: a new frontier for natural products. Chem Biol 5:R245–R249

Hojati Z, Milne C, Harvey B, Gordon L, Borg M, Flett F, Wilkinson B, Sidebottom PJ, Rudd BA, Hayes MA, Smith CP, Micklefield J (2002) Structure, biosynthetic origin, and engineered biosynthesis of calcium-dependent antibiotics from *Streptomyces coelicolor*. Chem Biol 9:1175–1187

Hopkins DW, MacNaughton SJ, O'Donnell AG (1991) A dispersion and differential centrifugation technique for representatively sampling microorganisms from soil. Soil Biol Biochem 23:217–225

Horan AC (1994) Aerobic actinomycetes: a continuing source of novel natural products. Biotechnology 26:3–30

Hu H, Ochi K (2001) Novel approach for improving the productivity of antibiotic-producing strains by inducing combined resistant mutations. Appl Environ Microbiol 67:1885–1892

Ikeda H, Nonomiya T, Usami M, Ohta T, Omura S (1999) Organization of the biosynthetic gene cluster for the polyketide anthelmintic macrolide avermectin in *Streptomyces avermitilis*. Proc Natl Acad Sci USA 96:9509–9514

Ikeda H, Ishikawa J, Hanamoto A, Shinose M, Kikuchi H, Shiba T, Sakaki Y, Hattori M, Omura S (2003) Complete genome sequence and comparative analysis of the industrial microorganism *Streptomyces avermitilis*. Nat Biotechnol 21:526–531

Kuhstoss S, Richardson MA, Rao RN (1991) Plasmid cloning vectors that integrate site-specifically in *Streptomyces spp.* Gene 97:143–146

Kurtböke I, Hayakawa M, Terekhova L, Okazaki T (2003) Selective isolation of rare actinomycetes. University of the Sunshine Coast/Queensland Complete Printing Services, Queensland, Australia

Langer M, Gabor EM, Liebeton K, Meurer G, Niehaus F, Schulze R, Eck J, Lorenz P (2006) Metagenomics: an inexhaustible access to nature's diversity. Biotechnol J 1:815–821

Liles MR, Manske BF, Bintrim SB, Handelsman J, Goodman RM (2003) A census of rRNA genes and linked genomic sequences within a soil metagenomic library. Appl Environ Microbiol 69:2684–2691

Lorenz P, Eck J (2005) Metagenomics and industrial applications. Nat Rev Microbiol 3:510–516

MacNeil IA, Tiong CL, Minor C, August PR, Grossman TH, Loiacono KA, Lynch BA, Phillips T, Narula S, Sundaramoorthi R, Tyler A, Aldredge T, Long H, Gilman M, Holt D, Osburne MS

(2001) Expression and isolation of antimicrobial small molecules from soil DNA libraries. J Mol Microbiol Biotechnol 3:301–308

Martin-Laurent F, Philippot L, Hallet S, Chaussod R, Germon JC, Soulas G, Catroux G (2001) DNA extraction from soils: old bias for new microbial diversity analysis methods. Appl Environ Microbiol 67:2354–2359

Martinez A, Kolvek SJ, Yip CL, Hopke J, Brown KA, MacNeil IA, Osburne MS (2004) Genetically modified bacterial strains and novel bacterial artificial chromosome shuttle vectors for constructing environmental libraries and detecting heterologous natural products in multiple expression hosts. Appl Environ Microbiol 70:2452–2463

Martinez A, Kolvek SJ, Hopke J, Yip CL, Osburne MS (2005) Environmental DNA fragment conferring early and increased sporulation and antibiotic production in *Streptomyces* species. Appl Environ Microbiol 71:1638–1641

McAlpine JB, Bachmann BO, Piraee M, Tremblay S, Alarco AM, Zazopoulos E, Farnet CM (2005) Microbial genomics as a guide to drug discovery and structural elucidation: ECO-02301, a novel antifungal agent, as an example. J Nat Prod 68:493–496

Miao V, Coeffet-Legal MF, Brian P, Brost R, Penn J, Whiting A, Martin S, Ford R, Parr I, Bouchard M, Silva CJ, Wrigley SK, Baltz RH (2005) Daptomycin biosynthesis in *Streptomyces roseosporus*: cloning and analysis of the gene cluster and revision of peptide stereochemistry. Microbiology 151:1507–1523

Miao V, Brost R, Chapple JP, She K, Coëffet-Le Gal M-F, Baltz RH (2006) The lipopeptide antibiotic A54145 biosynthetic gene cluster from *Streptomyces fradiae*. J Ind Microbiol Biotechnol 33:129–140

Miller DN, Bryant JE, Madsen EL, Ghiorse WC (1999) Evaluation and optimization of DNA extraction and purification procedures for soil and sediment samples. Appl Environ Microbiol 65:4715–4724

Mutka SC, Carney JR, Liu Y, Kennedy J (2006) Heterologous production of epothilone C and D in *Escherichia coli*. Biochemistry 45:1321–1330

Nalin R, Robe P, Van VT (2006) Method for extracting DNA from organisms. USA Patent 6,989,249

Neufeld JD, Mohn WW (2005) Unexpectedly high bacterial diversity in arctic tundra relative to boreal forest soils, revealed by serial analysis of ribosomal sequence tags. Appl Environ Microbiol 71:5710–5718

Ogram A, Sayler GS, Barkay T (1987) The extraction and purification of microbial DNA from sediments. J Microbiol Methods 7:57–66

Okamoto-Hosoya Y, Okamoto S, Ochi K (2003) Development of antibiotic-overproducing strains by site-directed mutagenesis of the *rpsL* gene in *Streptomyces lividans*. Appl Environ Microbiol 69:4256–4259

Otoguro M, Hayakawa M, Yamazaki T, Iimura Y (2001) An integrated method for the enrichment and selective isolation of *Actinokineospora spp.* in soil and plant litter. J Appl Microbiol 91:118–130

Pel J, Broemeling D, Chan I, Lu K, Mai L, Mercer G, Miao V, Tropini G, Davies J, Marziali A (2007) Direct extraction and purification of nucleic acids from environmental samples using 2-D electrophoretic concentration, LabAutomation 2007, The Association for Laboratory Automation Conference, Palm Springs, Jan 27-31 2007

Penn J, Li X, Whiting A, Wrigley SK, Latif M, Gibson T, Silva C, Davies JE, Miao V, Brian P, Baltz RH (2006) Heterologous production of daptomycin in *Streptomyces lividans*. J Ind Microbiol Biotechnol 33:121–128

Pfeifer B, Hu Z, Licari P, Khosla C (2002) Process and metabolic strategies for improved production of *Escherichia coli*-derived 6-deoxyerythronolide B. Appl Environ Microbiol 68:3287–3292

Pfeifer BA, Wang CC, Walsh CT, Khosla C (2003) Biosynthesis of yersiniabactin, a complex polyketide-nonribosomal peptide, using *Escherichia coli* as a heterologous host. Appl Environ Microbiol 69:6698–6702

Piepersberg W (1994) Pathway engineering in secondary metabolite-producing actinomycetes. Crit Rev Biotechnol 14:251–285

Quaiser A, Ochsenreiter T, Klenk HP, Kletzin A, Treusch AH, Meurer G, Eck J, Sensen CW, Schleper C (2002) First insight into the genome of an uncultivated crenarchaeote from soil. Environ Microbiol 4:603–611

Radomski CC, Seow KT, Warren RAJ, Yap WH (1998) Method for isolating xylanase gene sequences from soil DNA, compositions useful in such method and compositions obtained thereby. USA Patent 5,849,491, Dec 15, 1988

Riesenfeld CS, Goodman RM, Handelsman J (2004a) Uncultured soil bacteria are a reservoir of new antibiotic resistance genes. Environ Microbiol 6:981–989

Riesenfeld CS, Schloss PD, Handelsman J (2004b) Metagenomics: genomic analysis of microbial communities. Annu Rev Genet 38:525–552

Robe P, Nalin R, Capellano C, Vogel TM, Simonet P (2003) Extraction of DNA from soil. Eur J Soil Biol 39:183–190

Rondon MR, August PR, Bettermann AD, Brady SF, Grossman TH, Liles MR, Loiacono KA, Lynch BA, MacNeil IA, Minor C, Tiong CL, Gilman M, Osburne MS, Clardy J, Handelsman J, Goodman RM (2000) Cloning the soil metagenome: a strategy for accessing the genetic and functional diversity of uncultured microorganisms. Appl Environ Microbiol 66:2541–2547

Sandaa RA, Torsvik V, Enger O, Daae LF, Castberg T, Hahn D (1999) Analysis of bacterial communities in heavy metal-contaminated soils at different levels of resolution. FEMS Microbiol Ecol 30:237–251

Schloss PD, Handelsman J (2003) Biotechnological prospects from metagenomics. Curr Opin Biotechnol 14:303–310

Schwecke T, Aparicio JF, Molnar I, Konig A, Khaw LE, Haydock SF, Oliynyk M, Caffrey P, Cortes J, Lester JB et al. (1995) The biosynthetic gene cluster for the polyketide immunosuppressant rapamycin. Proc Natl Acad Sci USA 92:7839–7843

Seno E, Baltz R (1989) Structural organization and regulation of antibiotic biosynthesis and resistance genes in actinomycetes. In: Shipiro S (ed) Regulation of secondary metabolism in actinomycetes. CRC, Boca Raton, pp 1–48

Seow KT, Meurer G, Gerlitz M, WendtPienkowski E, Hutchinson CR, Davies J (1997) A study of iterative type II polyketide synthases, using bacterial genes cloned from soil DNA: a means to access and use genes from uncultured microorganisms. J Bacteriol 179:7360–7368

Short JM (2000) Screening for novel bioactivities. USA Patent 6,057,103

Short JM, Keller M (2004) Method for high-throughput screening of an environmental library. USA Patent 6,806,048

Silva CJ, Charman K, Sarroukh AM, Manohar RN, Penn J, Kau D, Brian P, Evans P, Englehardt J (2001) Screening for biological activity using calcium alginate encapsulated clones. In: International symposium on the biology of actinomycetes, Vancouver, August 5–9, 2001

Sosio M, Giusino F, Cappellano C, Bossi E, Puglia AM, Donadio S (2000) Artificial chromosomes for antibiotic-producing actinomycetes. Nat Biotechnol 18:343–345

Steffan RJ, Goksoyr J, Bej AK, Atlas RM (1988) Recovery of DNA from soils and sediments. Appl Environ Microbiol 54:2908–2915

Stein T (2005) *Bacillus subtilis* antibiotics: structures, syntheses and specific functions. Mol Microbiol 56:845–857

Streit WR, Schmitz RA (2004) Metagenomics–the key to the uncultured microbes. Curr Opin Microbiol 7:492–498

Strohl W (1997) Biotechnology of antibiotics, vol 82. Marcel Dekker, NY

Takahashi Y, Omura S (2003) Isolation of new actinomycete strains for the screening of new bioactive compounds. J Gen Appl Microbiol 49:141–154

Torsvik V, Goksoyr J, Daae FL (1990) High diversity in DNA of soil bacteria. Appl Environ Microbiol 56:782–787

Torsvik V, Daae FL, Sandaa RA, Ovreas L (1998) Novel techniques for analysing microbial diversity in natural and perturbed environments. J Biotechnol 64:53–62

Treusch AH, Kletzin A, Raddatz G, Ochsenreiter T, Quaiser A, Meurer G, Schuster SC, Schleper C (2004) Characterization of large-insert DNA libraries from soil for environmental genomic studies of Archaea. Environ Microbiol 6:970–980

Unwin J, Standage S, Alexander D, Hosted T Jr, Horan AC, Wellington EM (2004) Gene cluster in *Micromonospora echinospora* ATCC15835 for the biosynthesis of the gentamicin C complex. J Antibiot (Tokyo) 57:436–445

van Wageningen AM, Kirkpatrick PN, Williams DH, Harris BR, Kershaw JK, Lennard NJ, Jones M, Jones SJ, Solenberg PJ (1998) Sequencing and analysis of genes involved in the biosynthesis of a vancomycin group antibiotic. Chem Biol 5:155–162

Venter JC, Remington K, Heidelberg JF, Halpern AL, Rusch D, Eisen JA, Wu D, Paulsen I, Nelson KE, Nelson W, Fouts DE, Levy S, Knap AH, Lomas MW, Nealson K, White O, Peterson J, Hoffman J, Parsons R, Baden-Tillson H, Pfannkoch C, Rogers YH, Smith HO (2004) Environmental genome shotgun sequencing of the Sargasso Sea. Science 304:66–74

Verdine G (1996) The combinatorial chemistry of nature. Nature 384 suppl:11–13

Visca P, Imperi F, Lamont IL (2007) Pyoverdine siderophores: from biogenesis to biosignificance. Trends Microbiol 15:22–30

Wang GYS, Graziani E, Waters B, Pan WB, Li X, McDermott J, Meurer G, Saxena G, Andersen RJ, Davies J (2000) Novel natural products from soil DNA libraries in a streptomycete host. Org Lett 2:2401–2404

Ward AC, Bora N (2006) Diversity and biogeography of marine actinobacteria. Curr Opin Microbiol 9:279–286

Waters B, Davies J (1997) Amino acid variation in the GyrA subunit of bacteria potentially associated with natural resistance to fluoroquinolone antibiotics. Antimicrob Agents Chemother 41:2766–2769

Waters B, Miao V, Ho YW, Tong SK (1997) Method for isolation of biosynthesis genes for bioactive molecules. USA Patent 6,297,007

Watve MG, Tickoo R, Jog MM, Bhole BD (2001) How many antibiotics are produced by the genus *Streptomyces*? Arch Microbiol 176:386–390

Wawrik B, Kerkhof L, Zylstra GJ, Kukor JJ (2005) Identification of unique type II polyketide synthase genes in soil. Appl Environ Microbiol 71:2232–2238

Wenzel SC, Muller R (2005) Recent developments towards the heterologous expression of complex bacterial natural product biosynthetic pathways. Curr Opin Biotechnol 16:594–606

Wenzel SC, Gross F, Zhang Y, Fu J, Stewart AF, Muller R (2005a) Heterologous expression of a myxobacterial natural products assembly line in pseudomonads via red/ET recombineering. Chem Biol 12:349–356

Wenzel SC, Kunze B, Hofle G, Silakowski B, Scharfe M, Blocker H, Muller R (2005b) Structure and biosynthesis of myxochromides S1–3 in *Stigmatella aurantiaca*: evidence for an iterative bacterial type I polyketide synthase and for module skipping in nonribosomal peptide biosynthesis. Chembiochem 6:375–385

Williamson LL, Borlee BR, Schloss PD, Guan C, Allen HK, Handelsman J (2005) Intracellular screen to identify metagenomic clones that induce or inhibit a quorum-sensing biosensor. Appl Environ Microbiol 71:6335–6344

Yap WH, Li X, Soong TW, Davies JE (1996) Genetic diversity of soil microorganisms assessed by analysis of *hsp70* (*dnaK*) sequences. J Ind Microbiol Biotechnol 17:179–184

Yim G, de la Cruz F, Spiegelman GB, Davies J (2006a) Transcription modulation of *Salmonella enterica* serovar Typhimurium promoters by sub-MIC levels of rifampin. J Bacteriol 188:7988–7991

Yim G, Wang HH, Davies J (2006b) The truth about antibiotics. Int J Med Microbiol 296:163–170

Yun J, Ryu S (2005) Screening for novel enzymes from metagenome and SIGEX, as a way to improve it. Microb Cell Fact 4:8

Zazopoulos E, Huang K, Staffa A, Liu W, Bachmann BO, Nonaka K, Ahlert J, Thorson JS, Shen B, Farnet CM (2003) A genomics-guided approach for discovering and expressing cryptic metabolic pathways. Nat Biotechnol 21:187–190

Zengler K, Toledo G, Rappe M, Elkins J, Mathur EJ, Short JM, Keller M (2002) Cultivating the uncultured. Proc Natl Acad Sci USA 99:15681–15686

Zhou J, Bruns MA, Tiedje JM (1996) DNA recovery from soils of diverse composition. Appl Environ Microbiol 62:316–322

Ziermann R, Betlach MC (1999) Recombinant polyketide synthesis in *Streptomyces*: engineering of improved host strains. Biotechniques 26:106–110

Persisters, Biofilms, and the Problem of Cultivability

Kim Lewis

Contents

Abstract The majority of bacterial species in the environment remain uncultured, and accumulating evidence suggests that this cannot be explained by inadequate nutrient supply. Good recovery of environmental bacteria can be obtained by cultivation in situ in diffusion chambers, and this produces "domesticated" variants that can subsequently grow on synthetic media on Petri dishes. In many cases, growth of otherwise unculturable bacteria is observed on rich media in the presence of a cultivable helper organism. In the marine sediment environment, a considerable part of these uncultivable bacteria are found to depend on siderophores produced by their neighbors. The absence of an ability to induce the synthesis of their own siderophores when iron levels drop is puzzling. It seems that these observations point to a signaling mechanism for uncultivability – most bacterial species evolved to grow only in a familiar environment. The default mode of most bacterial life is then dormancy, and growth factors are required for resuscitation. The adaptive advantage of such a strategy may stem from the fact that rapidly propagating bacteria are highly vulnerable to toxic factors such as unfamiliar antibiotics. By contrast, dormant cells are tolerant to antibiotics. This is exemplified by specialized dormant persister cells which are formed in all studied cultivable bacteria. In organisms such as *E. coli* or *P. aeruginosa*, persisters are formed stochastically and make up a small part of the population. It is possible that in the absence of a growth factor, unculturable species enter en mass into a persister state.

K. Lewis
Antimicrobial Discovery Center and The Department of Biology,
Northeastern University,
360 Huntington Ave, Boston, MA, USA
e-mail: k.lewis@neu.edu

Microbiol Monogr, doi:10.1007/7171_2008_7

The list of unsolved problems in microbiology is short, and is topped by bacterial uncultivability:

- Unculturable bacteria ~1,900
- Persistent infections, tuberculosis ~1,900
- Persistence/multidrug tolerance 1,944
- Biofilm drug tolerance ~ 1,980

It seems that these apparently different phenomena studied by unrelated fields actually share a single underlying cause – the ability of bacteria to enter into a dormant state. This hypothesis will be critically examined in the present chapter.

Unculturable bacteria. Perhaps the greatest mystery in microbiology is that of "unculturable" bacteria. More than 99% of all species present in the environment refuse to grow on perfectly nutritious media in the lab. This defies our everyday experience with bacteria, which grow everywhere and contaminate everything and does not seem to make sense. Yet the "weeds" that we are so familiar with appear to be the rare exceptions to the common rule of unculturability. Let us start by examining the evidence.

The basic observation of unculturability is straightforward and is taught in some microbiology lab courses – an environmental sample such as soil is mixed with water and the concentration of bacterial cells is determined by microscopy count (usually with a fluorescent die), and by plating on a nutrient medium and counting the colonies. The results of this experiment, dating back more than 70 years (Butkevich 1932), is "The Great Plate Count Anomaly" (Staley and Konopka 1985) – the concentration of cells counted under the microscope is roughly 100 times higher than the estimate based on counting colonies. The majority of cells in the sample, therefore, do not grow into colonies.

Studies of environmental 16S rDNA showed that the anomaly is more than a gap in the total cell count – more than 99% of *species* from various environments are "unculturable" (Barer and Harwood 1999; Colwell and Grimes 2000; Giovannoni 2000; Grimes et al. 2000; Stackebrandt and Embley 2000). Uncultured organisms can be found in nearly every group within the Bacteria and Archaea, and several groups at the division level have been identified with no known cultivable representatives (Giovannoni et al. 1990; Torsvik et al. 1990; Ward et al. 1990; DeLong 1992; Fuhrman et al. 1992; Liesack and Stackebrandt 1992; Barns et al. 1994; Hugenholtz et al. 1998; Ravenschlag et al. 1999; Dojka et al. 2000; Rappe and Giovannoni 2003). We do not know the number of microbial species on this planet, but it is high, probably in the millions (Tiedje 1994; Allsopp et al. 1995). A single gram of soil contains about 10,000 bacterial species (Torsvik et al. 1990), and the estimate for a ton of soil is 4×10^6 species (Curtis et al. 2002). It is interesting to note that the estimated number of species increases (exponentially?) with time, and the latest calculations suggest 10^7 in a gram (rather than a ton) of soil (Gans et al. 2005). Up to 50% of the human oral microflora is unculturable (de Lillo et al. 2004; Munson et al. 2004), and 60–80% of the diverse gut flora consists of uncultivable species (Park et al. 2005). Unculturables seem to be thriving, but not on our Petri dishes. Popular explanations for the apparent uncultivability include very slow growth or a need for some unique nutrients (Rappe and Giovannoni 2003). While longer cultivation of environmental samples will somewhat increase the number of colonies on a Petri dish, patience alone has not been able to solve the problem of unculturability. The idea that unculturables require some special nutrients is hardly realistic. This would suggest that the majority of species choose not to use the most common compounds present in their environment – sugars and amino acids. The entire metabolic map is probably present in the conventional LB medium. It is important to stress that unculturable bacteria are found among all groups, including Pseudomonads and Actinomycetes famous for their metabolic versatility. One is forced to search for a nonobvious cause of uncultivability.

Recent research provides us with some important insights into the problem. A general method to grow unculturable bacteria from a variety of environments has been described (Kaeberlein et al. 2002; Lewis et al. 2002). The basic idea is to grow cells *in* their natural environment. Cells are taken from an environment such as marine sediment, diluted, mixed with agar, and sandwiched between two semipermeable membranes of a diffusion chamber. The chamber is then placed back in the same environment the sample came from, such as the sediment surface of a marine aquarium. The chamber permits the free diffusion of nutrients and other substances, while restricting the penetration of cells. After 1–2 weeks of incubation, colonies appear in the chamber. It is important to stress that simply placing a chamber in marine water or "dead" sediment does not work – the presence of an active sediment microbial community is essential. At the same time, placing a sample from the water column in marine water works, and resulted in the cultivation and sequencing of *Pelagibacter ubique*, the first representative of unculturable bacterial domain SAR11 and one of the most common marine bacteria previously known only by its 16S signature (Connon and Giovannoni 2002; Rappe et al. 2002; Zengler et al. 2002; Giovannoni et al. 2005). Similar success by growing in the environment has been reported for soil organisms – substantial recovery has been reported by inoculating a soil sample on the surface of a filter placed on top of the soil. This approach resulted in the growth of a representative of a highly diverse, but previously unculturable domain TM7 (Ferrari et al. 2005).

The marine sediment is very rich in organisms (10^9 cells g^{-1}, similar to soil (Epstein 1997)), but the recovery by standard cultivation methods is unusually poor (0.01–0.05%) (Llobet-Brossa et al. 1998; Cifuentes et al. 2000). This makes the sediment an ideal environment to work with unculturables because of the very low background of weeds. The diffusion chamber method improved the recovery ~300-fold. It follows from the above experiments that unculturables do grow rapidly in their natural environment. But why not on a Petri dish? We noticed that unculturable organisms will readily grow on nutrient medium on a Petri dish in vitro in the presence of other species from the same environment (Kaeberlein et al. 2002). This observation suggests that growth depends on substances (signals?) that indicate the presence of a familiar environment.

In a targeted approach to search for possible growth-promoting signals, a number of unculturable isolates that grow in the presence of a cultivable helper organism were obtained. It appeared that a simple pairwise reinoculation from a crowded plate seeded with a marine sediment sample readily leads to the isolation of culturable helper organisms that enable the growth, in close proximity, of unculturable bacteria. A particular pair consisting of a cultivable marine isolate of *M. luteus* and an uncultivable *Colwellia sp.* was examined further. The supernatant of *M. luteus* supported good growth of *Colwellia sp.*, and a bioassay-driven purification led to the identification of a number of siderophores, which proved to be responsible for the growth-promoting activity (Stewart et al. 2008). Interestingly, these growth experiments were performed on rich R2A medium optimized for culturing marine bacteria, and it was additionally supplemented with a mix of vitamins and metals, including iron. Siderophores increased the total cell count on R2A medium from this environment by nearly tenfold. It is possible that a mix containing a larger number of siderophores will improve recovery even more. Siderophores are nutrient-delivery molecules rather than signals. At the same time, it seems puzzling that a large number of species in the marine sediment would lack an ability to induce the production of siderophores when iron availability drops. Inducible synthesis of siderophores is a common theme in the microbial world, and it is very "cheap" to maintain inducible enzymes for siderophore biosynthesis. One interesting possibility is that a lack of siderophore synthesis genes provides bacteria with an ability to detect a familiar environment. If the correct siderophores from familiar neighbors are present, then iron is delivered and growth resumes; otherwise bacteria go into dormancy.

Another useful clue to the nature of uncultivability follows from the observation that unculturable bacteria can be readily "domesticated" by repeated reinoculation from chamber to chamber (Bollmann et al. 2007). This process produces readily cultivable variants. Apparently, the chamber serves as an intermediate between the natural environment and the Petri dish, selecting for domesticated progeny.

Dormancy – a default mode of bacterial life? The results of the co-culture and domestication experiments suggest that the majority of bacterial species evolved to grow in a familiar environment and "choose" not to grow when placed in a foreign medium. Signals from the familiar environment seem to be necessary to induce growth. The difference between weeds such as *E. coli* and *P. aeruginosa* and unculturables is then not their basic biochemical makeup, but a more sophisticated life strategy which results in growth arrest/dormancy in an unfamiliar environment. From this perspective, dormancy may be the default mode of most bacterial life.

In the subsequent sections, we will examine a well-documented case of bacterial dormancy, formation of persister cells. The ability of all bacterial species studied to enter a dormant state appears very similar to the growth arrest exhibited by unculturables in an unfamiliar environment. It is tempting to propose that an essentially similar dormancy program governs both persister cell formation and unculturability.

1 Discovery of Persister Cells

Persisters were described by Joseph Bigger in 1944 in one of the first studies of penicillin action (Bigger 1944). Bigger discovered that penicillin lysed a growing population of *Staphylococcus*, but plating this transparent solution on nutrient medium produced surviving colonies. In order to test whether these were mutants, the colonies were grown and treated with penicillin, and the new population again produced a small number of persisters surviving lysis. This experiment was repeated recently with *E. coli* as well as several different antibiotics and produced similar results (Keren et al. 2004a; Wiuff et al. 2005). But for a long time, Bigger's work was all but forgotten, a curiosity known to a few microbiologists. Harris Moyed picked up the problem in the 1980s and undertook a targeted search for persister genes (Moyed and Bertrand 1983; Moyed and Broderick 1986; Scherrer and Moyed 1988; Black et al. 1991; Black et al. 1994). He reasoned that treating a population of *E. coli* with ampicillin will select for mutants with increased production of persisters. After ampicillin application, cells were allowed to recover, and the enrichment process was repeated. This is different from the conventional selection for resistant mutants, where cells that can grow in the presence of antibiotic are favored. After testing for mutants that had the same MIC to ampicillin (thus not resistant), but survived better, several strains were obtained, and one of them was used to map the mutation to a *hipBA* locus. The mutant appeared to carry a mutation in the *hipA* gene, and this *hipA7* allelic strain was found to make 1% persisters in exponential cultures, about 1,000 times more than the wild type. Deletion of *hipBA* had no apparent effect on persister formation, suggesting that *hipA7* mutant (Korch et al. 2003) produced a pleiotropic artifact. Like Bigger's work before him, the studies of Moyed were largely forgotten.

The finding of persisters in biofilms rekindled an interest in this "curiosity", which appears to be responsible for a major part of recalcitrant human infectious diseases. We will summarize here what we currently know about the biology of persisters, and a more detailed account can be found in (Lewis 2007). But first, let us consider the difference between resistance of regular cells and drug tolerance of persisters. This will provide a useful framework for the subsequent discussion of persisters and their properties.

2 Biofilms and Persister Revival

Most bacterial species in the environment probably exist in the form of biofilms, bacterial communities that settle on a surface, and are covered by an exopolymer matrix (Costerton et al. 1995). Biofilms formed by pathogens are a major component of infectious diseases.

According to the Centers for Disease Control and Prevention (CDC), 65% of all infections in the developed countries are caused by biofilms, (Hall-Stoodley et al. 2004). These include common diseases such as child middle ear infection and gingivitis; infections of all known indwelling devices such as catheters, orthopedic prostheses, and heart valves; and the incurable disease of cystic fibrosis. Biofilms are produced by most if not all pathogens. *Pseudomonas aeruginosa* causing an incurable infection in cystic fibrosis patients (Singh et al. 2000) and *Staphylococcus aureus* and *Staphylococcus epidermidis* infecting indwelling devices (Mack et al. 2004) are probably the best-known biofilm-producing organisms. Biofilm infections are highly recalcitrant to antibiotic treatment. However, planktonic cells derived from these biofilms are in most cases fully susceptible to antibiotics. Importantly, biofilms do not actually grow in the presence of elevated levels of antibiotics, meaning they do not exhibit increased resistance as compared to planktonic cells (Lewis 2001a). But if biofilms are not resistant, how do they resist being killed? Biofilm resistance to killing has been one of the more elusive problems in microbiology, but the analysis of a simple dose–response experiment provided an unexpected insight into the puzzle (Brooun et al. 2000; Lewis 2001a; Spoering and Lewis 2001).

Most of the cells in a biofilm actually appear to be highly susceptible to a bactericidal agent such as a fluoroquinolone antibiotic or metal oxyanions, which can kill both rapidly dividing and slow- or nongrowing cells (Spoering and Lewis 2001; Harrison et al. 2005a; Harrison et al. 2005b) (Fig. 1). This is important, since cells in the biofilm are slow growing, and many are probably in stationary state. The experiment also revealed a small subpopulation of cells that remain alive irrespective of the concentration of the antibiotic. The level of these surviving persisters was even greater in the nongrowing stationary population. In a test tube, a stationary culture appears more tolerant than the biofilm. However, this situation is likely reversed in vivo. Antibiotic treatment will eliminate the bulk of both biofilm and planktonic cells, leaving intact persisters. At this point, the similarity with an in vitro experiment probably ends. The immune system will be able to mop up remaining planktonic persisters, just as it eliminates nongrowing cells of a population treated with a bacteriostatic antibiotic (Fig. 2). However, the biofilm matrix protects against immune cells (Leid et al. 2002; Jesaitis et al. 2003; Vuong et al. 2004), and its persisters will survive. After antibiotic concentration drops, persisters will repopulate the biofilm, which will shed off new planktonic cells, producing the relapsing biofilm infection (Lewis 2001a). The problem of biofilm resistance to "everything" largely defaults to understanding persisters.

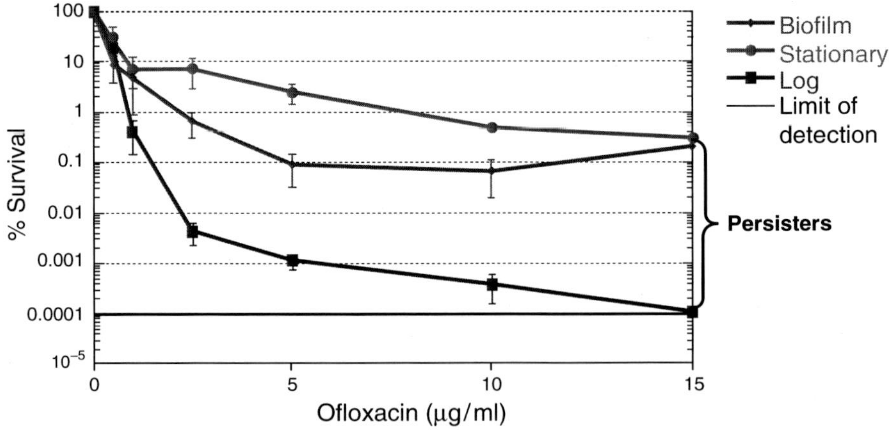

Fig. 1 Killing of logarithmic-phase, stationary-phase, and biofilm cultures of *P. aeruginosa* by ofloxacin. The limit of detection is indicated by the solid horizontal line

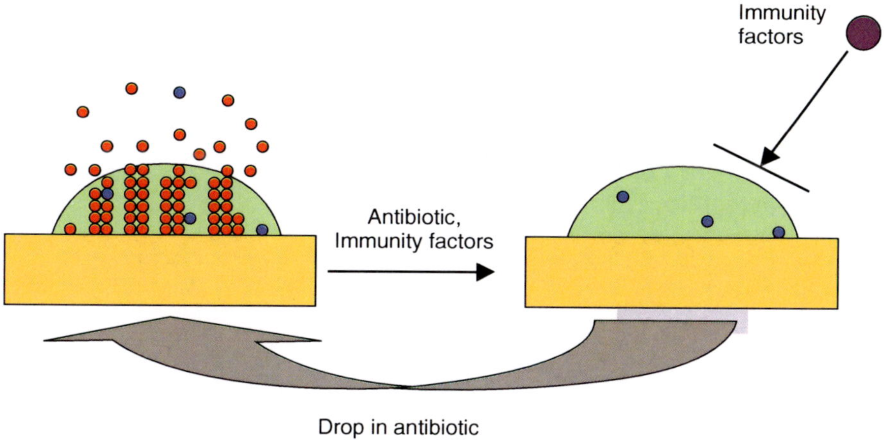

Fig. 2 Model of biofilm resistance based on persister survival. An initial treatment with antibiotic kills planktonic cells and the majority of biofilm cells. The immune system kills planktonic persisters, but the biofilm persister cells are protected from host defenses by the exopolysaccharide matrix. After the antibiotic concentration drops, persisters resurrect the biofilm and the infection relapses

The significance of persisters in sustaining a drug-tolerant infection rekindled the interest in this "curiosity".

In search of the mechanism of persister formation. Identification of the mechanism of persister formation presents a formidable challenge because of an apparent redundancy of persister genes. Thus, attempts to identify persister genes by screening transposon insertion libraries for either increased or decreased survival to antibiotics were not successful (Hu and Coates 2005; Spoering et al. 2006). The problem is exacerbated by a considerable variation in persister levels in parallel replicates (Wiuff et al. 2005), which becomes a major obstacle for screening large libraries, where it is responsible for an unmanageable level of false positives and negatives. An alternative method is to identify genes by screening/selecting a library cloned into an expression vector for gain of function. In this case, even a weak contributor to a multigene function can be identified when overexpressed. However, this approach is problematic as well, since overproduction of many proteins leads to misfolded toxic products that can stop cell growth and will create an artifact emulating a dormant state, as discussed above. It appears that standard approaches of molecular genetics are poorly suited to search for persister genes, which probably explains the slow pace of discovery in this area.

Another barrier to discovery has been a lack of approaches to isolate persister cells. The first method to isolate persisters was recently reported, based on simply sedimenting surviving cells from a culture lysed by ampicillin (Keren et al. 2004b). This method has its limitations – it requires a rapidly growing culture for ampicillin to lyse it, and the fraction of persisters in such a population is small, ~10-5. In *E. coli*, this necessitated the use of a *hipA7* strain overproducing persisters. In addition, these persisters are exposed to an antibiotic. These limitations notwithstanding, enough cells were collected to obtain a gene expression profile. The profile showed downregulation of proteins involved in energy production and nonessential functions such as flagellar synthesis, suggesting that persisters are dormant cells. This is consistent with the finding that persisters formed by a *hipA7* (high persistence) strain of *E. coli* are nongrowing (or slow growing) cells (Balaban et al. 2004). The profile also pointed to proteins that may be responsible for dormancy – RMF, a stationary state inhibitor of translation (Yoshida et al. 2002), SulA, an inhibitor of septation (Walker 1996), and toxin–antitoxin (TA) module elements RelBE, DinJ, and MazEF (Christensen and Gerdes 2003; Christensen et al. 2003). Homologs of TA modules are found on plasmids, where they constitute a maintenance mechanism (Hayes 2003). Typically, the toxin is a protein that inhibits an important cellular function such as translation or replication, and forms an inactive complex with the antitoxin. The toxin is stable, while the antitoxin is degradable. If a daughter cell does not receive a plasmid after segregation, the antitoxin level decreases owing to proteolysis, leaving a toxin that either kills the cell or inhibits propagation. TA modules are also commonly found on bacterial chromosomes, but their role is largely unknown. MazEF was proposed to serve as a programmed cell death mechanism (Sat et al. 2001). However, it was reported recently that MazF and an unrelated toxin RelE do not actually kill cells, but induce stasis by inhibiting translation, a condition that can be reversed by expression of corresponding antitoxins (Pedersen et al. 2002; Christensen et al. 2003).

Expression of RelE, a "toxin" that causes reversible stasis by inhibiting cleaving mRNA and inhibiting translation, strongly increased tolerance to antibiotics (Keren et al. 2004b). Expression of a toxin HipA increased tolerance as well (Falla and Chopra 1998; Correia et al. 2006; Korch and Hill 2006; Vazquez-Laslop et al. 2006). Interestingly, a bioinformatics analysis indicates that HipA is a member of the Tor family of kinases, which have been extensively studied in eukaryotes (Schmelzle and Hall 2000), but have not been previously identified in bacteria. HipA is indeed a kinase and it autophosphorylates on ser150, and site-directed mutagenesis replacing it or other conserved amino acids in the catalytic and Mg^{2+}-binding sites abolishes its ability to stop cell growth and confer drug tolerance (Correia et al. 2006). Knowing that HipA is a kinase provides an additional tool to search for the target, which is yet to be identified.

Deletion of potential candidates of persister genes noted above does not produce a discernible phenotype affecting persister production, possibly because of the high degree of redundancy of these elements. In *E. coli*, there are at least 10 toxin–antitoxin (TA) modules, and in *M. tuberculosis* more than 60 (Gerdes et al. 2005).

Several independent lines of evidence point to persister dormancy – lack of growth in the presence of antibiotics (by contrast to resistant mutants), downregulation of biosynthetic pathways, and an elegant demonstration of slow/no growth in persisters formed by the *E. coli hipA7* strain (Balaban et al. 2004). In the latter study, cells were placed in troughs of a multichannel chip, which restricts mobility and allows simultaneous videotaping of growth and division of many individual cells in the channels. The device also allowed flushing the medium, and application of ampicillin-caused lysis of cells. However, cells that did not lyse were those that had little growth preceding the application of ampicillin.

On the basis of these data, we reasoned that dormancy may be used to physically sort naïve persister cells from a wild-type population (Shah et al. 2006). Dormancy implies low levels of translation, which can then enable differential sorting based on expression of a detectable protein. In *E. coli* ASV, a degradable GFP is inserted into the chromosome in the l attachment site and expressed from the ribosomal *rrnB*P1 promoter, the activity of which is proportional to the rate of cell growth (Fig. 3). The half-life of degradable GFP is less than 1 h, and it should be effectively cleared from dormant cells. This would then enable sorting of dim persister cells. A logarithmically growing population of *E. coli* ASV was sorted with a high-speed cell sorter using forward light scatter, which allows detection of particles based on size. This enabled detection of cells irrespective of their level of fluorescence. Sorting by fluorescence showed that the population consisted of two strikingly different types of cells – a bright majority, and a small subpopulation of cells with no detectable fluorescence (Fig. 3). Fluorescent microscopy confirmed that the sorted bright cells were indeed bright green, while the dim ones had no detectable fluorescence. The dim cells were also smaller than the fluorescent cells, and in this regard resembled stationary-state cells. Sorted dim cells were exposed to a high level of ofloxacin, which rapidly kills both

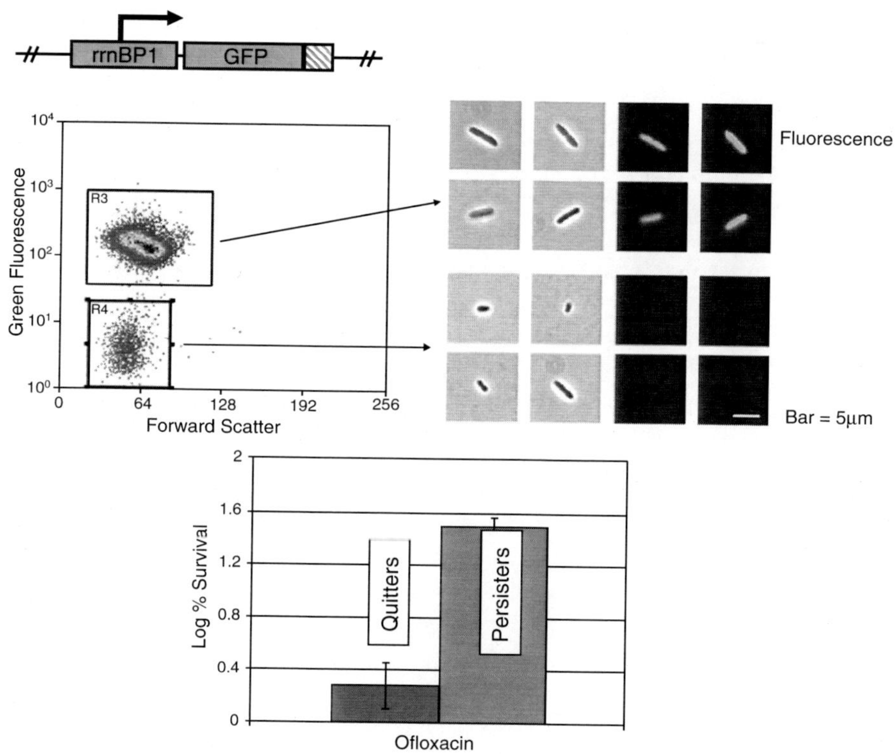

Fig. 3 Sorting of *E. coli* persister cells from a growing population. Cells carrying a degradable GFP under the control of a ribosomal promoter whose activity is proportional to the rate of growth were sorted, and the dim and bright cells were then exposed to an antibiotic to test their tolerance

growing and nongrowing normal cells, but has no effect on persisters. The majority of this subpopulation survived, as compared to a drastic drop in viability of the sorted bright cells. This experiment showed that the sorted dim cells are dormant persisters.

The sorting method provides a general approach to obtaining naïve persisters from a wild-type population of any species. These cells can then be used to obtain an expression profile and to study a variety of functions by biochemical methods.

Knowing that persisters are dormant cells bolsters the case for TA module involvement in persister formation. Indeed, TAs seem to be ideally suited for the task. Reversible action of "toxins" such as RelE and MazF, inhibition of important cellular functions by toxins capable of creating a dormant state, and the presence of TA modules in the chromosomes of all known free-living bacteria makes them attractive candidates for persister genes.

The multitude of proteins that can induce multidrug tolerance is reminiscent of the many MDR pumps responsible for multidrug resistance (MDR). *P. aeruginosa*, for example, contains genes coding for 15 MDR pumps belonging to the resistance-nodulation-cell division (RND) family alone, of which a single one, MexAb-OprM, is expressed at a high level under laboratory conditions (Li et al. 1995). Knockouts of most MDR genes produce no phenotype, while their overexpression produces a functional MDR pump (Lewis 2001b). It appears that microbial populations have evolved two complementary and highly redundant strategies to protect themselves from antimicrobials – multidrug efflux; and when this fails, multidrug tolerance of persister cells.

Persistence and uncultivability. What could be the downside of *not* growing in an otherwise nutritious, but unfamiliar, environment? A key to a possible solution to this problem lies in the simple fact that bacteria are most vulnerable to noxious factors when they rapidly divide. Thus a cell dividing in an unfamiliar but nutrient-rich environment exposes itself to an uncertain faith of being killed by an unfamiliar antibiotic to which it has no resistance. Similar to dormant persisters, cells of a dormant unculturable species would be well protected in an unfamiliar environment. Specific factors seem to be required for the growth of unculturables. Such a strategy is actually similar to the one used in the ultimate form of dormancy – spores of cultivable organisms whose germination strongly depends on environmental conditions. Spores will germinate on almost any medium that supports the growth of vegetative cells, but the efficiency of the process depends strongly on the presence of germination factors. For example, *B. subtilis* has a special receptor that detects the presence of the germination factor l-alanine (Setlow 2003). Unculturable bacteria seem to have taken this strategy one step further and developed a stringent requirement for cues that indicate the presence of a familiar environment in which they will commit to division. In the absence of such cues, they stay dormant.

References

Allsopp, D, Colwell, RR, Hawksworth, DL (1995) *Microbial diversity and ecosystem function*: Wallingford UK: CAB International

Balaban NQ, Merrin J, Chait R, Kowalik L, Leibler S (2004) Bacterial persistence as a phenotypic switch. *Science* 305:1622–1625

Barer MR, Harwood, CR (1999) Bacterial viability and culturability. Adv Microb Physiol 41:93–137

Barns SM, Fundyga RE, Jeffries MW, Pace NR (1994) Remarkable archaeal diversity detected in a Yellowstone National Park hot spring environment. Proc Natl Acad Sci U S A 91:1609–1613

Bigger JW (1944) Treatment of staphylococcal infections with penicillin. Lancet ii:497–500

Black DS, Kelly AJ, Mardis MJ, Moyed HS (1991) Structure and organization of *hip*, an operon that affects lethality due to inhibition of peptidoglycan or DNA synthesis. J Bacteriol 173:5732–5739

Black DS, Irwin B, Moyed HS (1994) Autoregulation of *hip*, an operon that affects lethality due to inhibition of peptidoglycan or DNA synthesis. J Bacteriol 176:4081–4091

Bollmann A, Lewis K, Epstein SS (2007) Incubation of environmental samples in a diffusion chamber increases the diversity of recovered isolates. Appl Environ Microbiol 73: 6386–6390

Brooun A, Liu S, Lewis K (2000) A dose-response study of antibiotic resistance in *Pseudomonas aeruginosa* biofilms. Antimicrob Agents Chemother 44:640–646

Butkevich VS (1932) Zür Methodik der bakterioloschen Meeresuntersuchungen und einige Angaben über die Verteilung der Bakterien im Wasser und in den Büden des Barents Meeres. Trans. Oceanogr. Inst. Moscow 2: 5–39 (in Russian with German summary) 2:5–39

Christensen SK, Gerdes K. (2003) RelE toxins from bacteria and Archaea cleave mRNAs on translating ribosomes, which are rescued by tmRNA. Mol Microbiol 48:1389–1400

Christensen SK, Pedersen K, Hansen FG, Gerdes K (2003) Toxin–antitoxin loci as stress-response-elements: ChpAK/MazF and ChpBK cleave translated RNAs and are counteracted by tmRNA. J Mol Biol 332:809–819

Cifuentes A, Anton J, Benlloch S, Donnelly A, Herbert RA, Rodriguez-Valera F (2000) Prokaryotic diversity in *Zostera noltii*-colonized marine sediments. Appl Environ Microbiol 66:1715–1719

Colwell RR, Grimes, D.J. (2000) Nonculturable microorganisms in the environment. American Society for Microbiology, Washington, DC

Connon SA, Giovannoni SJ (2002) High-throughput methods for culturing microorganisms in very-low-nutrient media yield diverse new marine isolates. Appl Environ Microbiol 68:3878–3885

Correia FF, D'Onofrio A, Rejtar T, Li L, Karger BL, Makarova K, Koonin EV, Lewis K (2006) Kinase activity of overexpressed HipA is required for growth arrest and multidrug tolerance in *Escherichia coli*. J Bacteriol 188:8360-8367.

Costerton JW, Lewandowski Z, Caldwell DE, Korber DR, Lappin-Scott HM (1995) Microbial biofilms. Annu Rev Microbiol 49:711–745

Curtis TP, Sloan WT, Scannell JW (2002) Estimating prokaryotic diversity and its limits. Proc Natl Acad Sci U S A 99:10494–10499

de Lillo A, Booth V, Kyriacou L, Weightman AJ, Wade WG (2004) Culture-independent identification of periodontitis-associated *Porphyromonas* and *Tannerella* populations by targeted molecular analysis. J Clin Microbiol 42:5523–5527

DeLong EF (1992) Archaea in coastal marine environments. Proc Natl Acad Sci U S A 89:5685–5689

Dojka MA, Harris JK, Pace NR (2000) Expanding the known diversity and environmental distribution of an uncultured phylogenetic division of bacteria. Appl Environ Microbiol 66:1617–1621

Epstein SS (1997) Microbial food webs in marine sediments. II. seasonal changes in trophic interactions in a sandy tidal flat Community. Microb Ecol 34:199–209

Falla TJ, Chopra I (1998) Joint tolerance to beta-lactam and fluoroquinolone antibiotics in *Escherichia coli* results from overexpression of *hipA*. Antimicrob Agents Chemother 42:3282–3284

Ferrari BC, Binnerup SJ, Gillings M (2005) Microcolony cultivation on a soil substrate membrane system selects for previously uncultured soil bacteria. Appl Environ Microbiol 71:8714–8720

Fuhrman JA, McCallum K, Davis AA (1992) Novel major archaebacterial group from marine plankton. Nature 356:148–149

Gans J, Wolinsky M, Dunbar J (2005) Computational improvements reveal great bacterial diversity and high metal toxicity in soil. Science 309:1387–1390

Gerdes K, Christensen SK, Lobner-Olesen A (2005) Prokaryotic toxin–antitoxin stress response loci. Nat Rev Microbiol 3:371–382

Giovannoni SJ, Britschgi TB, Moyer CL, Field KG (1990) Genetic diversity in Sargasso sea bacterioplankton. Nature 345:60–63

Giovannoni SJ, Tripp HJ, Givan S, Podar M, Vergin KL, Baptista D, Bibbs L, Eads J, Richardson TH, Noordewier M, Rappe MS, Short JM, Carrington JC, Mathur EJ (2005) Genome streamlining in a cosmopolitan oceanic bacterium. Science 309: 1242–1245

Giovannoni SJR (2000) Evolution, diversity and molecular ecology of marine prokaryotes. In: Kirchman D (ed) Microbial ecology of the oceans. Wiley-Liss, New York, pp 47–84

Grimes DJ, Mills AL, Nealson KH (2000) The importance of viable but nonculturable bacteria in biogeochemistry. In: Colwell RR and Grimes DJ (eds) Nonculturable microorganisms in the environment. ASM, Washington DC, pp 209–227

Hall-Stoodley L, Costerton JW, Stoodley P (2004) Bacterial biofilms: from the natural environment to infectious diseases. Nat Rev Microbiol 2:95–108

Harrison JJ, Ceri H, Roper NJ, Badry EA, Sproule KM, Turner RJ (2005a) Persister cells mediate tolerance to metal oxyanions in *Escherichia coli*. Microbiology 151:3181–3195

Harrison JJ, Turner RJ, Ceri H (2005b) Persister cells, the biofilm matrix and tolerance to metal cations in biofilm and planktonic *Pseudomonas aeruginosa*. Environ Microbiol 7:981–994

Hayes F (2003) Toxins-antitoxins: plasmid maintenance, programmed cell death, and cell cycle arrest. Science 301:1496–1499

Hu Y, Coates AR (2005) Transposon mutagenesis identifies genes which control antimicrobial drug tolerance in stationary-phase *Escherichia coli*. FEMS Microbiol Lett 243:117–124

Hugenholtz P, Goebel BM, Pace NR (1998) Impact of culture-independent studies on the emerging phylogenetic view of bacterial diversity. J Bacteriol 180: 4765–4774

Jesaitis AJ, Franklin MJ, Berglund D, Sasaki M, Lord CI, Bleazard JB, Duffy JE, Beyenal H, Lewandowski Z (2003) Compromised host defense on *Pseudomonas aeruginosa* biofilms: characterization of neutrophil and biofilm interactions. J Immunol 171:4329–4339

Kaeberlein T, Lewis K, Epstein SS (2002) Isolating "uncultivable" microorganisms in pure culture in a simulated natural environment. Science 296:1127–1129

Keren I, Kaldalu N, Spoering A, Wang Y, Lewis K (2004a) Persister cells and tolerance to antimicrobials. FEMS Microbiol Lett 230:13–18

Keren I, Shah D, Spoering A, Kaldalu N, Lewis K (2004b) Specialized persister cells and the mechanism of multidrug tolerance in Escherichia coli. J Bacteriol 186:8172–8180

Korch SB, Hill TM (2006) Ectopic overexpression of wild-type and mutant *hipA* genes in *Escherichia coli*: effects on macromolecular synthesis and persister formation. J Bacteriol 188:3826–3836

Korch SB, Henderson TA, Hill TM (2003) Characterization of the *hipA7* allele of *Escherichia coli* and evidence that high persistence is governed by (p)ppGpp synthesis. Mol Microbiol 50:1199–1213

Leid JG, Shirtliff ME, Costerton JW, Stoodley AP (2002) Human leukocytes adhere to, penetrate, and respond to *Staphylococcus aureus* biofilms. Infect Immun 70:6339 6345

Lewis K (2001a) Riddle of biofilm resistance. Antimicrob Agents Chemother 45:999–1007

Lewis K (2001b) In search of natural substrates and inhibitors of MDR pumps. J Mol Microbiol Biotechnol 3:247–254

Lewis K (2007) Persister cells, dormancy and infectious disease. Nat Rev Microbiol 5: 48–56

Lewis K, Salyers A, Taber H, Wax R (2002) Bacterial resistance to antimicrobials: mechanisms, genetics, medical practice and public health. New York, Marcel Dekker

Li XZ, Nikaido H, Poole K (1995) Role of *mexA-mexB-oprM* in antibiotic efflux in *Pseudomonas aeruginosa*. Antimicrob Agents Chemother 39:1948–1953

Liesack W, Stackebrandt E (1992) Occurrence of novel groups of the domain Bacteria as revealed by analysis of genetic material isolated from an Australian terrestrial environment. J Bacteriol 174:5072–5078

Llobet-Brossa E, Rossello-Mora R, Amann R (1998) Microbial community composition of wadden sea sediments as revealed by fluorescence *in situ* hybridization. Appl Environ Microbiol 64:2691–2696

Mack D, Becker P, Chatterjee I, Dobinsky S, Knobloch JK, Peters G, Rohde H, Herrmann M (2004) Mechanisms of biofilm formation in *Staphylococcus epidermidis* and *Staphylococcus aureus*: functional molecules, regulatory circuits, and adaptive responses. Int J Med Microbiol 294:203–212

Moyed HS, Bertrand KP (1983) *hipA*, a newly recognized gene of *Escherichia coli* K-12 that affects frequency of persistence after inhibition of murein synthesis. J Bacteriol 155:768–775

Moyed HS, Broderick SH (1986) Molecular cloning and expression of *hipA*, a gene of *Escherichia coli* K- 12 that affects frequency of persistence after inhibition of murein synthesis. J Bacteriol 166:399–403

Munson MA, Banerjee A, Watson TF, Wade WG (2004) Molecular analysis of the microflora associated with dental caries. J Clin Microbiol 42:3023–3029

Park HK, Shim SS, Kim SY, Park JH, Park SE, Kim HJ, Kang BC, Kim CM (2005) Molecular analysis of colonized bacteria in a human newborn infant gut. J Microbiol 43:345–353

Pedersen K, Christensen SK, Gerdes K (2002) Rapid induction and reversal of a bacteriostatic condition by controlled expression of toxins and antitoxins. Mol Microbiol 45:501–510

Rappe MS, Connon SA, Vergin KL, Giovannoni SJ (2002) Cultivation of the ubiquitous SAR11 marine bacterioplankton clade. Nature 418:630–633

Rappe MS, Giovannoni SJ (2003) The uncultured microbial majority. Annu Rev Microbiol 57:369–394

Ravenschlag K, Sahm K, Pernthaler J, Amann R (1999) High bacterial diversity in permanently cold marine sediments. Appl Environ Microbiol 65:3982–3989

Sat B, Hazan R, Fisher T, Khaner H, Glaser G, Engelberg-Kulka H (2001) Programmed cell death in *Escherichia coli*: some antibiotics can trigger *mazEF* lethality. J Bacteriol 183:2041–2045

Scherrer R, Moyed HS (1988) Conditional impairment of cell division and altered lethality in *hipA* mutants of *Escherichia coli* K-12. J Bacteriol 170:3321–3326

Schmelzle T, Hall MN (2000) TOR, a central controller of cell growth. Cell 103:253–262

Setlow P (2003) Spore germination. Curr Opin Microbiol 6:550–556

Shah DV, Zhang Z, Kurg K, Kaldalu N, Khodursky A, Lewis K (2006) Persisters: A distinct physiological state of *E. coli*. BMC Microbiol 6:53

Singh PK, Schaefer AL, Parsek MR, Moninger TO, Welsh MJ, Greenberg EP (2000) Quorum-sensing signals indicate that cystic fibrosis lungs are infected with bacterial biofilms. Nature 407:762–764

Spoering AL, Lewis K (2001) Biofilms and planktonic cells of *Pseudomonas aeruginosa* have similar resistance to killing by antimicrobials. J Bacteriol 183:6746–6751

Spoering AL, Vulic M, Lewis K (2006) GlpD and PlsB participate in persister cell formation in *Escherichia coli*. J Bacteriol 188:5136–5144

Stackebrandt E, Embley TM (2000) Diversity of uncultured microorganisms in the environment. In: Colwell RR and Grimes DJ (eds) Nonculturable microorganisms in the environment. ASM, Washington DC, pp 57–75

Staley JT, Konopka A (1985) Measurement of in situ activities of nonphotosynthetic microorganisms in aquatic and terrestrial habitats. Annu Rev Microbiol 39:321–346

Stewart E, D'Onofrio A, Witt K, Epstein S, Lewis K (2008) Identification of natural growth factors allows increased recovery of environmental bacterial isolates. Abstract N-042/0709. In *ASM General Meeting* ASM, Boston, MA

Tiedje JM (1994) Microbial diversity: of value to whom? ASM News 60:524–525

Torsvik V, Goksoyr J, Daae FL (1990) High diversity in DNA of soil bacteria. Appl Environ Microbiol 56:782–787

Vazquez-Laslop N, Lee H, Neyfakh AA (2006) Increased persistence in *Escherichia coli* caused by controlled expression of toxins or other unrelated proteins. J Bacteriol 188:3494–3497

Vuong C, Voyich JM, Fischer ER, Braughton KR, Whitney AR, DeLeo FR, Otto M (2004) Polysaccharide intercellular adhesin (PIA) protects *Staphylococcus epidermidis* against major components of the human innate immune system. Cell Microbiol 6:269–275

Walker GC (1996) The SOS response of *Escherichia coli*. In: *Escherichia coli* and *Samonella* cellular and molecular biology. ASM, Washington DC, pp 1400–1416

Ward DM, Weller R, Bateson MM (1990) 16S rRNA sequences reveal numerous uncultured microorganisms in a natural community. Nature 345:63–65

Wiuff C, Zappala RM, Regoes RR, Garner KN, Baquero F, Levin BR (2005) Phenotypic tolerance: antibiotic enrichment of noninherited resistance in bacterial populations. Antimicrob Agents Chemother 49:1483–1494

Yoshida H, Maki Y, Kato H, Fujisawa H, Izutsu K, Wada C, Wada A (2002) The ribosome modulation factor (RMF) binding site on the 100S ribosome of *Escherichia coli*. J Biochem 132:983–989

Zengler K, Toledo G, Rappe M, Elkins J, Mathur EJ, Short JM, Keller M (2002) Cultivating the uncultured. Proc Natl Acad Sci U S A 99:15681–15686

Taking the Concept to the Limit: Uncultivable Bacteria and Astrobiology

Kenneth Nealson

Contents

Abstract Cultivating microbes is enough of a challenge when one knows they are there: the microbial world is rife with examples of bugs that are clearly growing in nature, yet seem to be uncultivable despite our best efforts. While many of these clearly cultivable "uncultivables" defy our efforts, as a group these microbes may provide some valuable clues and lessons for those who would venture into the world of the search for extraterrestrial life. That is to say, careful consideration of how one might recognize and/or cultivate ET organisms might yield some insights into the more realistic question of how to find and cultivate life in earthly habitats – after all, our ability to cultivate microbes that we can clearly see are capable of growth has been disappointing to say the least. Thus, an open-minded approach to why our own earthly efforts have yielded such limited success might well be used to formulate a strategy for the cultivation of ET life, and perhaps more importantly, to improve our efforts to cultivate life from our own planet. And incidentally, should we be so fortunate as to ever obtain uncontaminated extraterrestrial samples, we will be ready!

K. Nealson
Wrigley Professor of Geobiology, University of Southern California,
Los Angeles, CA, 90089, USA
e-mail: knealson@usc.edu

Microbiol Monogr, doi:10.1007/7171_2009_3

1 Cultivation of Microbes

One of the historical tenets of microbiology has been that the organisms on which we work should be cultivated and studied as pure cultures. Indeed the proper description of a new species has traditionally required the inclusion of many properties that can only be gained by the study of the growing microbe in the laboratory (Sneath 1992). This being said, we now know that the majority of recognizable microbes seen microscopically (Whitman et al. 1998) as well as the vast majority of ribotypes (Olsen et al. 1994; Pace 1997; Woese 2004) seen as molecular signatures from virtually any environment have not been obtained in pure, or even mixed, culture. Take for instance the case of the light organ population of bacteria from the luminous fish *Photoblepharon palpabratus*, which I studied for several years. The bacteria in the light organ of this fish, which supply the light that it uses for a variety of purposes (Morin et al. 1975) reach densities of 10^9 ml^{-1} and have been shown to be growing very well in situ, with a doubling time of about 10 h (Haygood et al. 1984): like many other such "uncultivable" exosymbionts, they are regularly expelled from the host to the outside seawater (Nealson et al. 1984). Despite being obviously cultivable, however, these bacteria have so far defied all efforts to cultivate them outside their host. In fact, these bacteria, if encountered in a sample of seawater by a microbiologist, would be placed in the category of uncultivable! One thus imagines that much of this is a matter of semantics – surely the fish does not regard them as uncultivable! Such situations are common in symbioses of various kinds, ranging from gut symbionts in termites (Breznak and Leadbetter 2002; Stingl et al. 2005), to extracellular symbioses of many types, to intracellular insect/microbial symbioses (O'Neill et al. 1992; Brownlie and O'Neill 2006; Moran and Degnan 2006) and all the way to intracellular organelles like mitochondria and chloroplasts (Margulis 1981). In addition, there are abundant and complex metabolic symbioses between microbes, perhaps far more than have been suspected. These range from the consortia of bacteria and archaea working together for anaerobic methane oxidation (Boetius et al. 2000), to the recently described *Nanoarchaeum symbioticum* (Huber et al. 2002; Waters et al. 2003), and the elusive ARMAN-types: ultrasmall bugs from acidic environments (Baker et al. 2006) as well as the more common syntrophic symbioses based on interspecies hydrogen transfer (Fenchel and Finlay 1995; Schink and Stams 2002). What many, perhaps most, seem to have in common is that they are, by our definition, uncultivable!

But why are so many of our beloved microbes seemingly unwilling to grow for us? The estimates for the part of the populations that can be cultivated in most environments range from a few percent to 0.1% or less (Whitman et al. 1998). Perhaps the conditions are not right – we do not understand their physiology, biochemistry, or nutrient needs. Perhaps they are lacking a key metabolic partner: i.e., they are incapable of growing without the specific input (nutrient, trace element, regulatory element, etc.) of another organism(s). Perhaps they grow extremely slowly, so that we, in our impatience, incorrectly score them as negatives even though they are in fact growing. Perhaps under some conditions they slow down their metabolism to "maintenance" levels – avoiding growth and its complications: patiently waiting for things to get better. Of course, it could be a combination of any of the above and to this end, I will make a speculative proposal later, suggesting that many microbes prefer to grow slowly, working in consortia to take advantage of limiting resources, and we as microbiologists have neglected this part of the microbial world at the expense of the less abundant, but more easily seen, fast-growing phenotypes. Before that however, let us consider the fundamentals involved.

2 Growth as a Phenotype

How do we score growth? In the world of microbiology, we are spoiled – many of the organisms we use for teaching, and upon which we work are capable of rapid growth, and we have developed many techniques for studying this rapid growth via a variety of approaches (DNA, RNA, protein, total biomass, etc.). While the versatility of these organisms seems at first glance quite impressive, it is, in fact rather limited: most are heterotrophs that can utilize a few carbon sources, and even fewer electron acceptors. Thus, designing a suitable medium for such organisms is, for the most part, not a daunting task. An energy source to "eat," an electron acceptor to breathe, a carbon source to supply the molecular skeletons needed to make a new cell (often the same as the energy source), and the required nutrients (N, S, P, etc.) usually suffice. As a measure of success, we visualize a density increase in liquid medium, often indicating 10^7 (or more) cells per ml. To make it more daunting for the microbes, we may use a solid medium of agar or some alternative composition (Shungu et al. 1983), hoping to visualize single colonies of the bugs in question.

Cultivation becomes more challenging as one moves to more limiting, and/or hostile environments, or does it? In so-called extreme environments, the resident microbes have exquisitely adapted to the conditions that are sufficiently harsh as to make it difficult for the nongrowing microbes to "hang around and wait," and to some degree, it is a bit easier to coax microbes into the growth phase, as demonstrated, in our case, by the work of Takai and colleagues (Inagaki et al. 2003, 2004; Takai et al. 2002, 2003a, b, 2004). Many common microbial habitats have a kind of extremism I have referred to as "nutritional extremophily" (Nealson 2001; Nealson and Cox 2002; Nealson et al. 2002; Nealson and Berelson 2003) – favoring microbes that grow lithotrophically, using a wide range of inorganic electron donors, autotrophically incorporating CO_2, and heterotrophically utilizing a wide array of organic and inorganic electron acceptors, and phototrophically, utilizing anoxygenic or oxygenic photosynthesis, or perhaps even rhodopsin-based systems (Beja et al. 2000; Venter et al. 2004; Sharma et al. 2006; Gomez-Consarnau et al. 2007). Given these differences, plus expected differences in environmental temperature, pH, salinity, one expects that on any one medium, one might allow growth of only a small percentage of what one sees microscopically, as demonstrated by the hypothetical situations shown in Table 1, where suggested numbers have been assigned to each variable. For example, it is proposed that only 10% of the microscopically visible population is able to grow at the temperature chosen for study,

Table 1 Effect of multiple variables on apparent "cultivability"

Variable	Percent of bugs tolerant of a given range for the variable[a]	Cumulative percentage of "cultivable" bugs
Temperature	10 (50)	10 (50)
Salinity	20 (50)	2 (25)
pH	20 (50)	0.4 (12.5)
C – source	10 (50)	0.04 (6.25)
Electron acceptor	50 (50)	0.02 (3.12)
Electron donor	25 (50)	0.005 (1.56)

[a]This estimate is for demonstration only – we simply assign a value to each of the variables. In any given environment it could be higher or lower. For comparison, a second column is included in which each variable allows 50% of the population to succeed with growth

and so on. A second set of numbers shows the same impact but with much more conservative numbers: here it is considered that 50% of the microbes will tolerate each variable chosen. For practical reasons (cost of materials, space, manpower, etc.) it is common to do viable counts on a single medium at a single temperature, pH, and salinity. Assuming a wide distribution of metabolically diverse microbes in a sample, we see that one quickly moves to 1% or less of the microbes growing on any single medium that is used. Even if one assumes that each variable will allow 50% of the organisms to grow (Table 1, parentheses), after only six variables, one sees less than 2% population capable of growth. Surely, as more variables are added, it is inevitable that one moves to 0.1% or less quite easily – it should be no surprise. While the numbers to be assigned to any of these (and other) variables must be regarded as unknown, the point is easily made – we should expect low cultivability, rather than be surprised by it! However, whether it is proper to refer to the nongrowing fraction of microbes as noncultivable is debatable, to say the least.

Of course it is not so simple – more than one of these critical variables may apply to a given strain, so that the total will not be a simple multiplier of all the probabilities. This being said, the point remains that one predicts low success in cultivation if one (or only a few) media and conditions are used. One also predicts that if ten different media and/or conditions are used, they will yield ten different populations. Given this kind of thinking, it would be surprising if our primitive methods did better. Assuming a high diversity, we should always expect to see some growth, and seldom expect to see clonal-like high percentage cultivability in natural situations.

For the most part, microbiologists have been trained to believe that the most successful microbes are those that grow the fastest and/or reach the highest cell numbers. Is this necessarily the case in the environment? It is easy to imagine that it is not. For example, for the sake of discussion, I propose here the "straw-man" hypothesis that under limiting conditions, the most successful microbes are those that "know when to (i.e., have the metabolic and regulatory abilities to) stop growing." After all, if conditions are limiting, the last thing a microbe should want is a copy of itself – a sister cell with exactly the same requirements. It would be far better to incorporate another microbe that might consume metabolic waste products, and perhaps supply something in return! Such cells might be no more uncultivable than the luminous bacteria discussed above, but may require careful conditions to tease them out of their slow- or nongrowth metabolic state.

Such ideas are consistent with reports of very slow growth in deep subsurface environments – places where the influx of nutrients is extremely low and slow, being a function of low-nutrient groundwater flow, and the availability of lithospheric energy sources (Kerr 2001, 2002; D'Hondt et al. 2002, 2004). For example, D'Hondt et al. (2002) have calculated that average growth rates of deep subsurface microbes are in the range of 100s to 1000s of years. If so, it renders the use of growth as a measure of viability a nonuseful exercise! That is, one should ask how these microbes manage to survive, and what can be measured to assess their viability. Recent studies have used metabolic methods to establish that many (perhaps most) of the microbes are in fact alive, but growing at low rates that would be virtually cryptic to our usual methods (Schippers et al. 2005). Taken together, the concept of noncultivable may well be more a statement about our own abilities (or inabilities) than about the organisms we are trying to cultivate.

3 Life's Signatures

Composition constitutes one of the most recognizable signatures of life. The particular combination of elements found in our life is remarkably constant for earthly life (C, H, N, P, O, S, K, Fe), and virtually unknown in the mineral world (Nealson and Conrad 1999; Dorn et al. 2003; Dorn 2005). In a similar vein, the distribution and abundance of monomers (amino acids and fatty acids, in particular) defies thermodynamic explanation, and is a strong indicator of a living process (Dorn et al. 2003; Dorn 2005). Of course the abundance of repeated polymeric molecules is unmistakably a sign of life, but often much harder to measure.

Another approach to life's signatures involves the kinetic signatures via enzymatic catalysis: i.e., the consumption of nutrients and production of products at rates inconsistent with simple chemistry. Such catalyses occur with organic and/or inorganic molecules, and often result in gradients of electron acceptors such as oxygen, nitrate, sulfate, CO_2, and others (Nealson and Conrad 1999;2001 Nealson and Berelson 2003; Nealson). These gradients are formed as a result of metabolic processes that consume reactants or produce products at rates far greater than nutrients can be resupplied or waste products can be dissipated by diffusion. Thus, in nearly all sediments, where convective mixing is absent or minimal, oxygen depletion occurs due to aerobic metabolism, resulting in an oxygen gradient. Such gradients can be rapidly dissipated upon death of the organisms due to diffusion, but may well leave recognizable traces, even when the gradients are gone (Nealson and Berelson 2009).

4 Detecting Life When No Growth is Occurring

Assuming that life is cellular, then the search for life might well begin with a visible search for structures that appear to be cells, followed by analysis of such structures for compositional signatures [elemental and monomer (amino acid and fatty acid) content]. For living organisms, growing or not, such signals should be useful indicators of life. Recently this approach was taken for some "nanobacteria" (Baker et al. 2006) and "nanoarchaea" (Huber et al. 2002; Waters et al. 2003), yielding convincing evidence that these small microbe-like structures were indeed part of the living world. As opposed to many reports of purported nan(n)obacteria in the literature, which have been supported only by visual images, these two examples provide evidence for life at what must be near the lower size limit. Not only did the elemental composition fit with life as we know it, but the ultrastructural analyses revealed cellular structures, and membranes.

5 Kinetic Signatures of Life

As for the kinetic signatures, slow, or nongrowing life could well be cryptic. If metabolism is sufficiently slow, then all needs dealing with nutrient uptake and/or waste disposal might be supplied by diffusion, a rationale for developing a very slow growth strategy. If metabolism can be adjusted to be so slow that wastes diffuse out, and nutrients diffuse in, faster than they are needed, then gradients will essentially disappear. To this end, one might imagine that early life was so inefficient that it was not diffusion limited – a distinct advantage in the early evolution of metabolism!

There are many places on our own planet where the above metabolic strategies might prove useful, such as very extreme environments, the deep subsurface, and sites of extreme nutrient limitation. Such efforts can be "ground-truthed" using other approaches such as molecular probes, DNA analyses, etc. to confirm the results.

6 Searching for Life in Extraterrestrial Samples

Uncontaminated extraterrestrial samples, other than samples of moon rocks brought back by the Apollo astronauts have never been available for study. Should they ever be obtained one might well like to pursue a strategy like the one described above. That is, first looking for structures reminiscent of cells, and then examining these structures for their chemical content (inorganics, organic monomers, etc.). One would not necessarily be looking for an elemental composition characteristic of earthly life, but rather a composition that is "unusual" or inconsistent with the mineralogy or geochemistry of the remainder of the sample. Similar statements could be made regarding the monomers present. Are there monomers (amino acids, sugars, fatty acids, etc.), and do these fit with a chemical explanation? For example, the

Murchison meteorite samples were consistent with the results of the Miller–Urey experiment, suggesting that the monomers arose by an abiotic mechanism – the monomers (amino acids, fatty acids, etc.) in standard Earthly bacteria defy explanation by abiotic pathways (Dorn 2005)! What would one surmise if the distribution of monomers was different from that predicted by abiotic synthesis? Certainly not that the sample is alive: rather one would now have identified a sample whose composition needs a credible explanation.

7 Uncultivable ET Life!

Should one be so fortunate as to see evidence for "life" in an ET sample, then it would be desirable to have a clearly delineated strategy for cultivation. Given our modest success with earthly samples, one should not hope for too much, but the following strategy might be useful.

1. Knowing the origin of the sample, one should take into account probable electron donors and acceptors that might be, or have been, available.
2. Knowing the elemental and monomer composition of the purported life forms, try to supply all elements and other components as needed in the growth medium.
3. Try a variety of solid and liquid media.
4. Watch for increases in cell number or mass of the sample.
5. Carefully measure the metabolic activity of the samples, looking for evidence of metabolism, even in the absence of growth.

Of course, as mentioned above, if the sample has great metabolic diversity, something in it might grow. One would then be challenged to characterize the growing creature, to develop methods for detecting it in detail, and ultimately to establish that it was not a contaminant.

Since the chance of finding life seems so meager, and the chance that it would actually grow for us earthlings seems even less, why do we bother? For one thing, it would be incredibly exciting to find life from another place, and then get the pleasure of characterizing and explaining it! Secondly, some of the approaches we develop for this work might very well enhance our knowledge of our own planet: would not it be interesting if approaches developed for ET life were to reveal new life forms in extreme niches on Earth?

8 A Proposal for the Origin and Early Evolution of Life: Food for Thought

I end with some modest thoughts about how life might have begun, what that means on Earth (and perhaps elsewhere), and why we so often fail in our attempts to cultivate life on our own Planet. The basic features of my proposal are:

1. Life arose to take advantage of the most abundant energy and carbon source(s) available (hydrogen, formate, ammonium, reduced metals) for energy, CO_2 for carbon: i.e., lithotrophic and autotrophic life abounded, and the world was ruled by inorganic energy harvesting.
2. Life survived by living in low-energy environments with metabolic rates sufficiently low that they obviated the need for motility and took advantage of natural rates of diffusion: survival was key, and the ability to grow rapidly was not an important invention at this time. In fact, rapid growth may have been a substantial disadvantage, with sister cells competing for slowly diffusing, limited resources.
3. Under such conditions, self-sustaining metabolic reactions allowed life to evolve and adapt to the conditions, and "invent" the ability to rapidly synthesize both monomers and polymeric catalysts. As metabolic products accumulated (methane and acetate, for example), microbes of various sorts "learned" how to utilize these products, leading to syntrophic relationships of the type(s) that are so common today in the anaerobic world.
4. Organisms learned how to replicate and grow, but this was not a strongly selected trait. Keeping in balance with the nutrients provided by their geological environment, and now, their biological neighbors was the way of the world.
5. As the world became slowly more oxidized, new "metabolic inventions" occurred, and heterotrophic metabolism became abundant. Up to this point, given that microbes are not inherently predatory, competition was for chemical resources, and cooperation was a favored strategy.
6. About 2 Ga (billion years) ago the first eukaryotes appeared. Assuming that they were able to prey on bacteria, the world changed drastically. At this point, the ability to grow must have become a major important feature of the microbial world. Given that the predatory eukaryotes were oxygen dependent, two strategies developed. First, to maintain the slow-growth phenotype, but reside in anaerobic environments where eukaryotic predation was not abundant; or second, to grow faster than predation could remove them.

The above scenario, while a bit fanciful, may have some important aspects to it with regard to searching for life. It has hidden within it, the notion that there may be environments on Earth where such strategies still exist – slow growth, competition for chemical resources, and syntrophic cooperation among the community members. How do we find such microbes? We would propose looking for physical and/or chemical structures – sites of the concentration of elements or monomers that are not easily explained by chemistry alone. Such sites could be the starting point for characterizing life, be it growing, nongrowing, dead, or even fossilized. If it is growing, then a careful assessment of how it grows may provide a clue to its cultivation, but remember the lessons of *Photoblepharon* and other symbionts: the chance of failure is great if the approach is cultivation!

References

Baker BJ, Tyson GW et al (2006) Lineages of acidophilic Archaea revealed by community genomic analysis. Science 314:1933–1935

Beja O, Aravind L et al (2000) Bacterial rhodopsin: evidence for a new type of phototrophy in the sea. Science 289:1902–1906

Boetius A, Ravenschlag K et al (2000) A marine microbial consortium apparently mediating anaerobic oxidation of methane. Nature 407:623–626

Breznak JA, Leadbetter JR (2002) Termite gut spirochetes. In: Dworkin M, Falkow S, Rosenberg E, Schliefer KH, Stackebrandt E (eds) The Prokaryotes – an evolving electronic resource for the microbiological community. Springer, New York

Brownlie JC, O'Neill SL (2006) *Wolbachia* genomes: insights into an intracellular lifestyle. Curr Biol 15:R507–R509

D'Hondt S, Rutherford S, Spivack AJ (2002) Metabolic activity of subsurface life in deep-sea sediments. Science 295:2067–2070

D'Hondt S, Joergensen BB et al (2004) Distributions of microbial activities in deep subseafloor sediments. Science 306:2216–2221

Dorn ED (2005) Universal biosignatures for the detection of life. Computer Sciences. Pasadena, California Institute of Technology, California, p 123

Dorn ED, McDonald GD et al (2003) Principal component analysis and neural networks for detection of amino acid biosignatures. Icarus 166(2):403–409

Fenchel T, Finlay BJ (1995) Ecology and evolution in anoxic worlds. Oxford University Press, Oxford

Gomez-Consarnau L, Gonzalez JM et al (2007) Light stimulates growth of proteorhodopsin-containing marine *Flavobacteria*. Nature 445:210–213

Haygood MG, Tebo BM et al (1984) Luminous bacteria of a monocentrid fish *(Monocentris-Japonicus)* and 2 anomalopid fishes *(Photoblepharon-Palpebratus and Kryptophanaron-Alfredi)* – population sizes and growth within the light organs, and rates of release into the seawater. Mar Biol 78(3):249–254

Huber H, Hohn MI et al (2002) A new phylum of Archaea represented by a nanosized hyperthermophilic symbiont. Nature 417:63–67

Inagaki F, Takai K et al (2003) *Sulfurimonas autotrophica* gen. nov., sp nov., a novel sulfur-oxidizing epsilon-proteobacterium isolated from hydrothermal sediments in the mid-Okinawa trough. Int J Syst Evol Micr 53:1801–1805

Inagaki F, Takai K et al (2004) Sulfrovum lithotrophicum gen.nov., sp. nov., a novel sulfur-oxidizing epsilon-proteobacterium isolated from hydrothermal sediments from the mid-Okinawa trough. Int J Syst Evol Microbiol 54:1477–1487

Kerr RA (2001) Life – potential, slow, or long dead? Science 294:1820–1821

Kerr RA (2002) Deep life in the slow, slow lane. Science 296:1056–1058

Margulis L (1981) Symbiosis in cell evolution. Freeman, San Francisco

Moran NA, Degnan PH (2006) Functional genomics of *Buchnera* and the ecology of aphid hosts. Mol Ecol 15:1251–1261

Morin JG, Harrington A et al (1975) Light for all reasons – versatility in behavioral repertoire of flashlight fish. Science 190(4209):74–76

Nealson K (2001) Searching for life in the universe: lessons from the Earth. Ann N Y Acad Sci 950:241–258

Nealson K, Berelson W (2009) Sedimentary habitats. In: Schaechter M (ed) Desk encyclopedia of microbiology. Academic, London, p 800

Nealson K, Berelson W (2003) Layered microbial communities and the search for life in the universe. Geomicrobiol J 20:451–462

Nealson KH, Conrad PG (1999) Life: past, present and future. Phil Trans R Soc Lond B 354:1–17

Nealson KH, Cox BL (2002) Microbial metal-ion reduction and Mars: extraterrestrial expectations? Curr Opin Microbiol 5:296–300
Nealson KH, Haygood MG et al (1984) Contribution by symbiotically luminous fishes to the occurrence and bioluminescence of luminous bacteria in seawater. Microb Ecol 10(1):69–77
Nealson KH, Tsapin A, Storrie-Lombardi M (2002) Searching for life in the Universe: unconventional methods for an unconventional problem. Int Microbiol 5:223–230
O'Neill SL, Giordano R et al (1992) 16S rRNA phylogenetic analysis of the bacterial endosymbionts associated with cytoplasmic incompatibility in insects. Proc Natl Acad Sci U S A 89:2699–2672
Olsen GJ, Woese CR et al (1994) The winds of (evolutionary) change: breathing new life into microbiology. J Bacteriol 176:1–6
Pace NR (1997) A molecular view of microbial diversity and the biosphere in a Yellowstone hot spring. Science 276:734–740
Schink B, Stams AJM (2002) Syntrophism among prokaryotes. In: Dworkin M (ed) The prokaryotes: an evolving electronic resource for the microbiological community. Springer, New York
Schippers A, Neretin LN et al (2005) Prokaryotic cells of the deep sub-seafloor biosphere identified as living bacteria. Nature 433:861–864
Sharma AK, Spudich JL et al (2006) Microbial rhodopsins: functional versatility and genetic mobility. Trends Microbiol 14:463–469
Shungu D, Valiant M et al (1983) Gelrite as an agar substitute in bacteriological media. Appl Environ Microbiol 46(4):840–845
Sneath PHA (1992) International code of nomenclature of bacteria: bacteriological code, 1990 revision. American Society Microbiology, Washington, DC
Stingl U, Radek R et al (2005) *Endomicrobia*: cytoplasmic symbionts of termite gut protozoa from a separate phylum of prokaryotes. Appl Environ Microbiol 71:1473–1479
Takai K, Hirayama H, Sakihama Y, Inagaki F, Yamato Y, Horikoshi K (2002) Isolation and metabolic characteristics of previously uncultured members of the order Aquifales in a subsurface gold mine. Appl Environ Microbiol 68:3046–3054
Takai K, Inagaki F et al (2003a) Isolation and phylogenetic diversity of members of previously uncultivated epsilon-proteobacteria in deep-sea hydrothermal fields. FEMS Microbiol Lett 218(1):167–174
Takai K, Kobayashi H et al (2003b) *Deferribacter desulfuricans* sp nov., a novel sulfur-, nitrate- and arsenate-reducing thermophile isolated from a deep-sea hydrothermal vent. Int J Syst Evol Microbiol 53:839–846
Takai K, Nealson KH et al (2004) *Hydrogenimonas thermophila* gen. nov., sp nov., a novel thermophilic, hydrogen-oxidizing chemolithoautotroph within the epsilon-proteobacteria, isolated from a black smoker in a Central Indian Ridge hydrothermal field. Int J Syst Evol Microbiol 54:25–32
Venter JC, Remington K et al (2004) Environmental genome shotgun sequencing of the Sargasso sea. Science 304(5667):66–74
Waters E, Hohn MJ et al (2003) The genome of *Nanoarchaeum equitans*: insights into early archael evolution and derived parasitism. Proc Natl Acad Sci U S A 100:12984–12988
Whitman BW, Coleman DC, Wiebe WJ (1998) Prokaryotes: the unseen majority. Proc Natl Acad Sci U S A 95:6578–6583
Woese CR (2004) A new biology for a new century. Micorbiol Mol Biol Rev 68:173–186

Index

M

N

O

P

MIX
Papier aus verantwortungsvollen Quellen
Paper from responsible sources
FSC® C105338

Printed by Books on Demand, Germany